Edward R. Tufte

The Visual Display
of Quantitative Information

Graphics Press · Cheshire, Connecticut

Contents

For my parents
Edward E. Tufte and Virginia James Tufte

Acknowledgments

I am indebted to many for their advice and assistance with this book.

For leave and research support during several academic years, the Center for Advanced Study in the Behavioral Sciences, the John Simon Guggenheim Foundation, the Woodrow Wilson School of Princeton University, and Yale University.

For providing access to their superb collections, the Bibliothèque Nationale and the Bibliothèque de l'École Nationale des Ponts et Chaussées in Paris, and, at Yale University, the Historical Medical Library and the Beinecke Rare Book and Manuscript Library.

For helping me appreciate the practicalities in the production of statistical graphics, several members of the art department at the *New York Times* and my students in Graphic Design and in the Department of Statistics at Yale.

For assistance in establishing Graphics Press, Peter B. Cooper, Earle E. Jacobs, Jr., and Trudy Putsche.

For design and artwork, Howard I. Gralla and Minoru Niijima.

For their help and hospitality in Paris during my work on the Minard drawings, Michel Balinski, Jean Dubout, André Jammes, and Claudine Kleb.

For providing examples and for suggesting improvements in the manuscript, James Beniger, Inge Druckrey, Timothy Gregoire, Joanna Hitchcock, Joseph LaPalombara, Kathryn Scholle, Stephen Stigler, Howard Wainer, and Ellen Woodbury.

For their reviews of the manuscript and for their inspiration and encouragement through all the years of this enterprise, Frederick Mosteller and John W. Tukey.

June 1982
Cheshire, Connecticut

Introduction

Data graphics visually display measured quantities by means of the combined use of points, lines, a coordinate system, numbers, symbols, words, shading, and color.

The use of abstract, non-representational pictures to show numbers is a surprisingly recent invention, perhaps because of the diversity of skills required—the visual-artistic, empirical-statistical, and mathematical. It was not until 1750–1800 that statistical graphics—length and area to show quantity, time-series, scatterplots, and multivariate displays—were invented, long after such triumphs of mathematical ingenuity as logarithms, Cartesian coordinates, the calculus, and the basics of probability theory. The remarkable William Playfair (1759–1823) developed or improved upon nearly all the fundamental graphical designs, seeking to replace conventional tables of numbers with the systematic visual representations of his "linear arithmetic."

Modern data graphics can do much more than simply substitute for small statistical tables. At their best, graphics are instruments for reasoning about quantitative information. Often the most effective way to describe, explore, and summarize a set of numbers—even a very large set—is to look at pictures of those numbers. Furthermore, of all methods for analyzing and communicating statistical information, well-designed data graphics are usually the simplest and at the same time the most powerful.

The first part of this book reviews the graphical practice of the two centuries since Playfair. The reader will, I hope, rejoice in the graphical glories shown in Chapter 1 and then condemn the lapses and lost opportunities exhibited in Chapter 2. Chapter 3, on graphical integrity and sophistication, seeks to account for these differences in quality of graphical design.

The second part of the book provides a language for discussing graphics and a practical theory of data graphics. Applying to most visual displays of quantitative information, the theory leads to changes and improvements in design, suggests why some graphics might be better than others, and generates new types of graphics. The emphasis is on maximizing principles, empirical measures of graphical performance, and the sequential improvement of graphics through revision and editing. Insights into graphical design are to be gained, I believe, from theories of what makes for excellence in art, architecture, and prose.

This is a book about the design of statistical graphics and, as such, it is concerned both with design and with statistics. But it is also about how to communicate information through the simultaneous presentation of words, numbers, and pictures. The design of statistical graphics is a universal matter—like mathematics—and is not tied to the unique features of a particular language. The descriptive concepts (a vocabulary for graphics) and the principles advanced apply to most designs. I have at times provided evidence about the scope of these ideas, by showing how frequently a principle applies to (a random sample of) news and scientific graphics.

Each year, the world over, somewhere between 900 billion (9×10^{11}) and 2 trillion (2×10^{12}) images of statistical graphics are printed. The principles of this book apply to most of those graphics. Some of the suggested changes are small, but others are substantial, with consequences for hundreds of billions of printed pages.

But I hope also that the book has consequences for the viewers and makers of those images—that they will never view or create statistical graphics the same way again. That is in part because we are about to see, collected here, so many wonderful drawings, those of Playfair, of Minard, of Marey, and, nowadays, of the computer.

Most of all, then, this book is a celebration of data graphics.

PART I

Graphical Practice

I *Graphical Excellence*

Excellence in statistical graphics consists of complex ideas communicated with clarity, precision, and efficiency. Graphical displays should

- show the data

- induce the viewer to think about the substance rather than about methodology, graphic design, the technology of graphic production, or something else

- avoid distorting what the data have to say

- present many numbers in a small space

- make large data sets coherent

- encourage the eye to compare different pieces of data

- reveal the data at several levels of detail, from a broad overview to the fine structure

- serve a reasonably clear purpose: description, exploration, tabulation, or decoration

- be closely integrated with the statistical and verbal descriptions of a data set.

Graphics *reveal* data. Indeed graphics can be more precise and revealing than conventional statistical computations. Consider Anscombe's quartet: all four of these data sets are described by exactly the same linear model (at least until the residuals are examined).

I		II		III		IV	
X	Y	X	Y	X	Y	X	Y
10.0	8.04	10.0	9.14	10.0	7.46	8.0	6.58
8.0	6.95	8.0	8.14	8.0	6.77	8.0	5.76
13.0	7.58	13.0	8.74	13.0	12.74	8.0	7.71
9.0	8.81	9.0	8.77	9.0	7.11	8.0	8.84
11.0	8.33	11.0	9.26	11.0	7.81	8.0	8.47
14.0	9.96	14.0	8.10	14.0	8.84	8.0	7.04
6.0	7.24	6.0	6.13	6.0	6.08	8.0	5.25
4.0	4.26	4.0	3.10	4.0	5.39	19.0	12.50
12.0	10.84	12.0	9.13	12.0	8.15	8.0	5.56
7.0	4.82	7.0	7.26	7.0	6.42	8.0	7.91
5.0	5.68	5.0	4.74	5.0	5.73	8.0	6.89

$N = 11$

mean of X's $= 9.0$

mean of Y's $= 7.5$

equation of regression line: $Y = 3 + 0.5X$

standard error of estimate of slope $= 0.118$

$t = 4.24$

sum of squares $X - \overline{X} = 110.0$

regression sum of squares $= 27.50$

residual sum of squares of $Y = 13.75$

correlation coefficient $= .82$

$r^2 = .67$

And yet how they differ, as the graphical display of the data makes vividly clear:

F. J. Anscombe, "Graphs in Statistical Analysis," *American Statistician*, 27 (February 1973), 17–21.

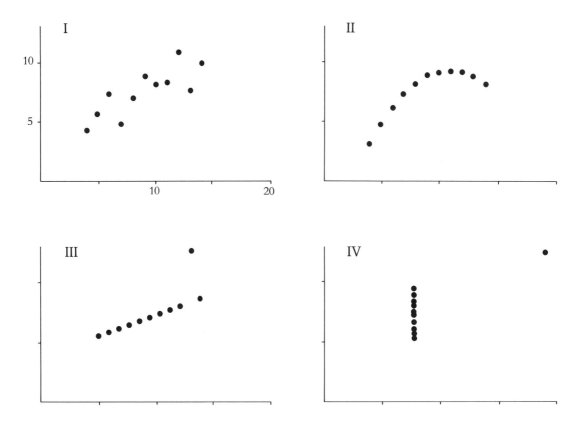

And likewise a graphic easily reveals point A, a wildshot observation that will dominate standard statistical calculations. Note that point A hides in the marginal distribution but shows up as clearly exceptional in the bivariate scatter.

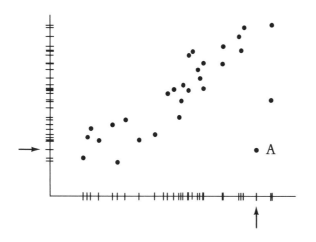

Stephen S. Brier and Stephen E. Fienberg, "Recent Econometric Modelling of Crime and Punishment: Support for the Deterrence Hypothesis?" in Stephen E. Fienberg and Albert J. Reiss, Jr., eds., *Indicators of Crime and Criminal Justice: Quantitative Studies* (Washington, D.C., 1980), p. 89.

Of course, statistical graphics, just like statistical calculations, are only as good as what goes into them. An ill-specified or preposterous model or a puny data set cannot be rescued by a graphic (or by calculation), no matter how clever or fancy. A silly theory means a silly graphic:

Edward R. Dewey and Edwin F. Dakin, *Cycles: The Science of Prediction* (New York, 1947), p. 144.

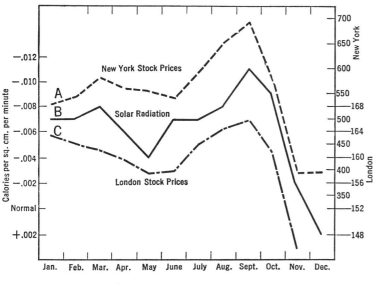

SOLAR RADIATION AND STOCK PRICES

A. New York stock prices (Barron's average). B. Solar Radiation, inverted, and C. London stock prices, all by months, 1929 (after Garcia-Mata and Shaffner).

Let us turn to the practice of graphical excellence, the efficient communication of complex quantitative ideas. Excellence, nearly always of a multivariate sort, is illustrated here for fundamental graphical designs: data maps, time-series, space-time narrative designs, and relational graphics. These examples serve several purposes, providing a set of high-quality graphics that can be discussed (and sometimes even redrawn) in constructing a theory of data graphics, helping to demonstrate a descriptive terminology, and telling in brief about the history of graphical development. Most of all, we will be able to see just how good statistical graphics can be.

Data Maps

These six maps report the age-adjusted death rate from various types of cancer for the 3,056 counties of the United States. Each map portrays some 21,000 numbers.[1] Only a picture can carry such a volume of data in such a small space. Furthermore, all that data, thanks to the graphic, can be thought about in many different ways at many different levels of analysis—ranging from the contemplation of general overall patterns to the detection of very fine county-by-county detail. To take just a few examples, look at the

- high death rates from cancer in the northeast part of the country and around the Great Lakes

- low rates in an east-west band across the middle of the country

- higher rates for men than for women in the south, particularly Louisiana (cancers probably caused by occupational exposure, from working with asbestos in shipyards)

- unusual hot spots, including northern Minnesota and a few counties in Iowa and Nebraska along the Missouri River

- differences in types of cancer by region (for example, the high rates of stomach cancer in the north-central part of the country —probably the result of the consumption of smoked fish by Scandinavians)

- rates in areas where you have lived.

The maps provide many leads into the causes—and avoidance— of cancer. For example, the authors report:

> In certain situations . . . the unusual experience of a county warrants further investigation. For example, Salem County, New Jersey, leads the nation in bladder cancer mortality among white men. We attribute this excess risk to occupational exposures, since about 25 percent of the employed persons in this county work in the chemical industry, particularly the manufacturing of organic chemicals, which may cause bladder tumors. After the finding was communicated to New Jersey health officials, a company in the area reported that at least 330 workers in a single plant had developed bladder cancer during the last 50 years. It is urgent that surveys of cancer risk and programs in cancer control be initiated among workers and former workers in this area.[2]

[1] Each county's rate is located in two dimensions and, further, at least four numbers would be necessary to reconstruct the size and shape of each county. This yields 7×3,056 entries in a data matrix sufficient to reproduce a map.

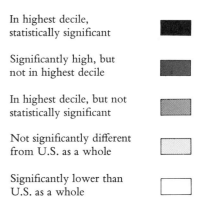

In highest decile, statistically significant

Significantly high, but not in highest decile

In highest decile, but not statistically significant

Not significantly different from U.S. as a whole

Significantly lower than U.S. as a whole

[2] Robert Hoover, Thomas J. Mason, Frank W. McKay, and Joseph F. Fraumeni, Jr., "Cancer by County: New Resource for Etiologic Clues," *Science*, 189 (September 19, 1975), 1006.

Maps from *Atlas of Cancer Mortality for U.S. Counties: 1950–1969*, by Thomas J. Mason, Frank W. McKay, Robert Hoover, William J. Blot, and Joseph F. Fraumeni, Jr. (Washington, D.C.: Public Health Service, National Institutes of Health, 1975). The six maps shown here were redesigned and redrawn by Lawrence Fahey and Edward Tufte.

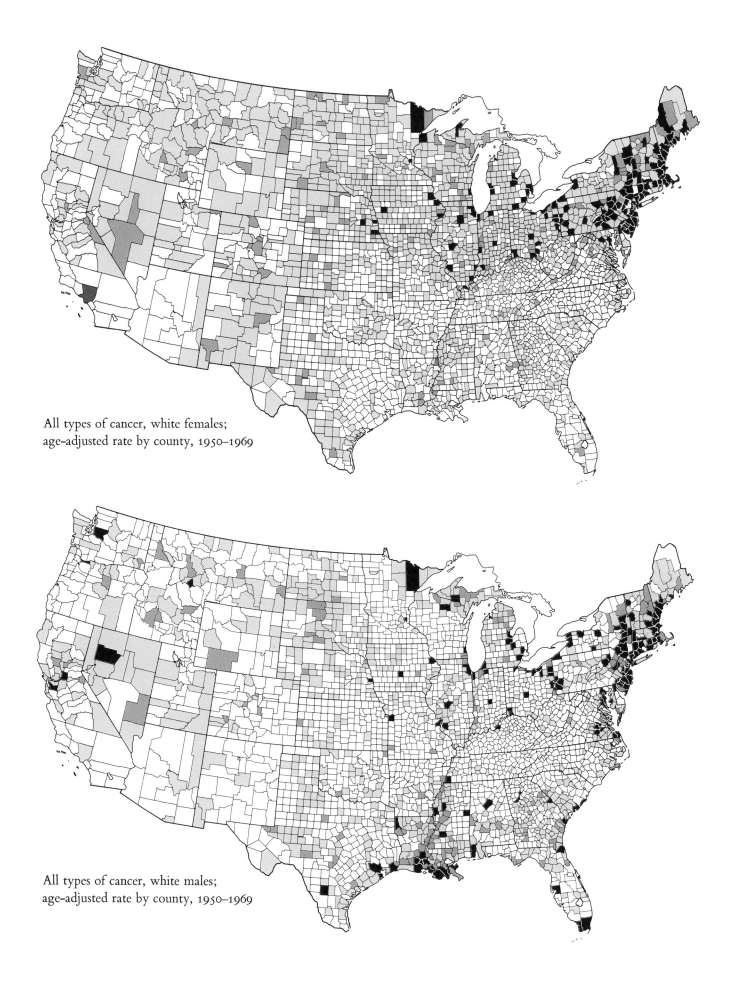

All types of cancer, white females;
age-adjusted rate by county, 1950–1969

All types of cancer, white males;
age-adjusted rate by county, 1950–1969

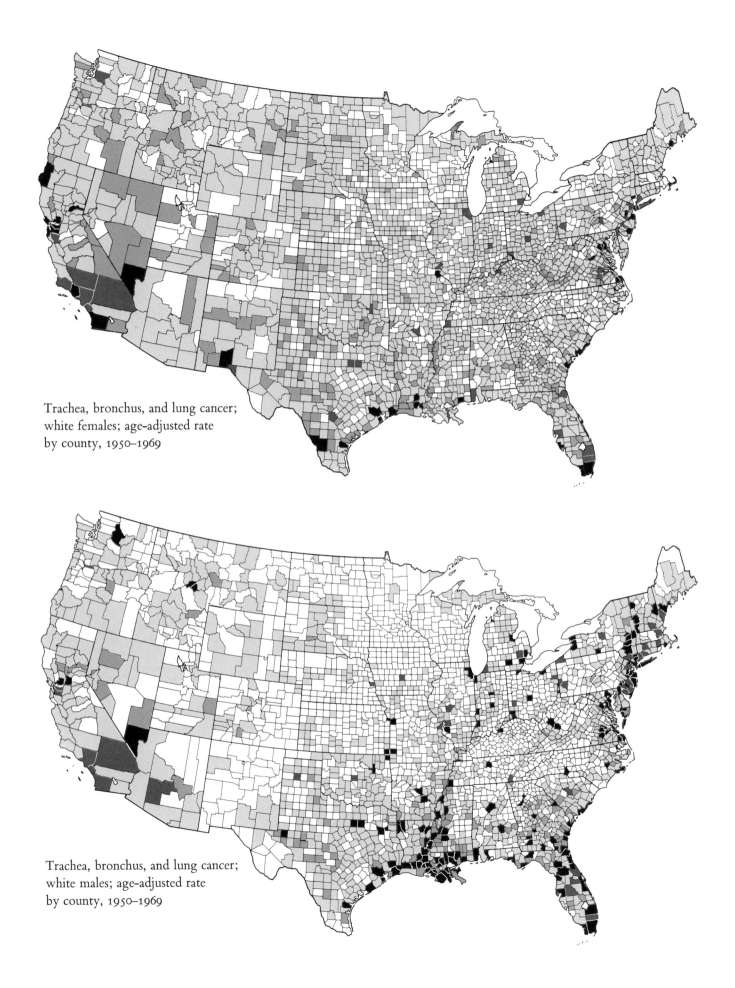

Trachea, bronchus, and lung cancer;
white females; age-adjusted rate
by county, 1950–1969

Trachea, bronchus, and lung cancer;
white males; age-adjusted rate
by county, 1950–1969

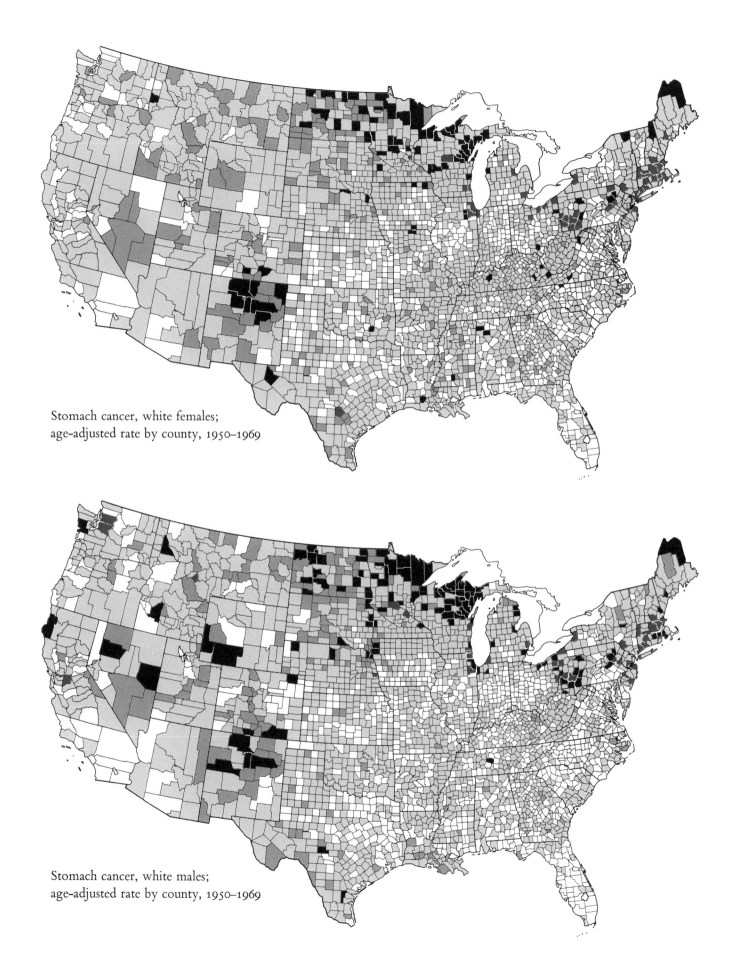

Stomach cancer, white females;
age-adjusted rate by county, 1950–1969

Stomach cancer, white males;
age-adjusted rate by county, 1950–1969

The maps repay careful study. Notice how quickly and naturally our attention has been directed toward exploring the substantive content of the data rather than toward questions of methodology and technique. Nonetheless the maps do have their flaws. They wrongly equate the visual importance of each county with its geographic area rather than with the number of people living in the county (or the number of cancer deaths). Our visual impression of the data is entangled with the circumstance of geographic boundaries, shapes, and areas—the chronic problem afflicting shaded-in-area designs of such "blot maps" or "patch maps."

A further shortcoming, a defect of data rather than graphical composition, is that the maps are founded on a suspect data source, death certificate reports on the cause of death. These reports fall under the influence of diagnostic fashions prevailing among doctors and coroners in particular places and times, a troublesome adulterant of the evidence purporting to describe the already sometimes ambiguous matter of the exact bodily site of the primary cancer. Thus part of the regional clustering seen on the maps, as well as some of the hot spots, may reflect varying diagnostic customs and fads along with the actual differences in cancer rates between areas.

Data maps have a curious history. It was not until the seventeenth century that the combination of cartographic and statistical skills required to construct the data map came together, fully 5,000 years after the first geographic maps were drawn on clay tablets. And many highly sophisticated geographic maps were produced centuries before the first map containing any statistical material was drawn.[3] For example, a detailed map with a full grid was engraved during the eleventh century A.D. in China. The Yü Chi Thu (Map of the Tracks of Yü the Great) shown here is described by Joseph Needham as the

> . . . most remarkable cartographic work of its age in any culture, carved in stone in +1137 but probably dating from before +1100. The scale of the grid is 100 *li* to the division. The coastal outline is relatively firm and the precision of the network of river systems extraordinary. The size of the original, which is now in the Pei Lin Museum at Sian, is about 3 feet square. The name of the geographer is not known. . . . Anyone who compares this map with the contemporary productions of European religious cosmography cannot but be amazed at the extent to which Chinese geography was at that time ahead of the West. . . . There was nothing like it in Europe till the Escorial MS. map of about +1550. . . .[4]

[3] Data maps are usually described as "thematic maps" in cartography. For a thorough account, see Arthur H. Robinson, *Early Thematic Mapping in the History of Cartography* (Chicago, 1982). On the history of statistical graphics, see H. Gray Funkhouser, "Historical Development of the Graphical Representation of Statistical Data," *Osiris*, 3 (November 1937), 269–404; and James R. Beniger and Dorothy L. Robyn, "Quantitative Graphics in Statistics: A Brief History," *American Statistician*, 32 (February 1978), 1–11.

[4] Joseph Needham, *Science and Civilisation in China* (Cambridge, 1959), vol. 3, 546–547.

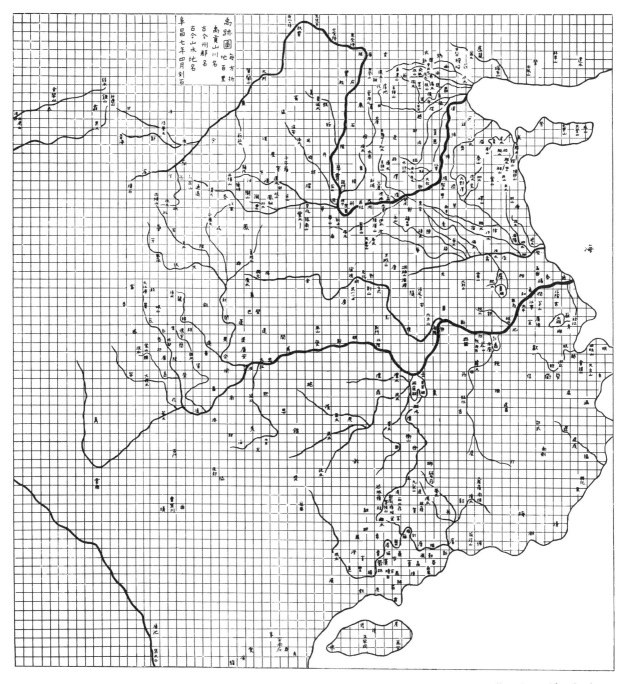

E. Chavannes, "Les Deux Plus Anciens
Spécimens de la Cartographie Chinoise,"
*Bulletin de l'École Française de l'Extrême
Orient*, 3 (1903), 1–35, Carte B.

Ecce formulam, vſum, atque

ſtructuram Tabularum Ptolomæi, cum quibuſdam locis, in quibus ſtudioſus Geographiæ ſe ſatis exercere poteſt.

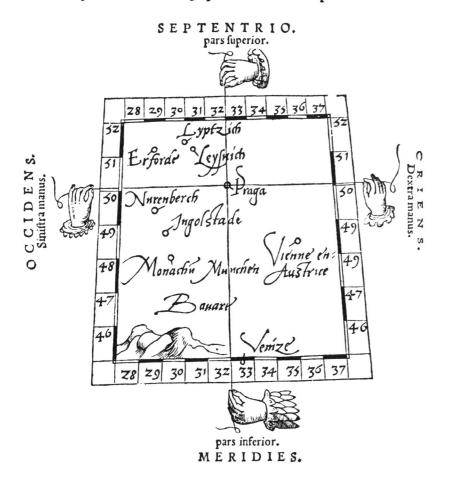

The 1546 edition of *Cosmographia* by Petrus Apianus contained examples of map design that show how very close European cartography by that time had come to achieving statistical graphicacy, even approaching the bivariate scatterplot. But, according to the historical record, no one had yet made the quantitative abstraction of placing a measured quantity on the map's surface at the intersection of the two threads instead of the name of a city, let alone the more difficult abstraction of replacing latitude and longitude with some other dimensions, such as time and money. Indeed, it was not until 1786 that the first economic time-series was plotted.

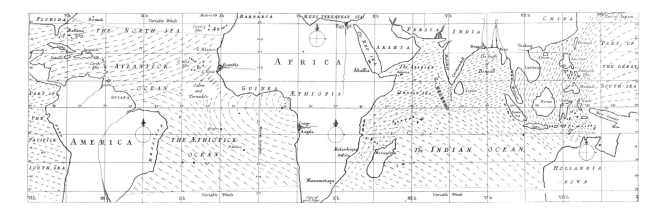

One of the first data maps was Edmond Halley's 1686 chart showing trade winds and monsoons on a world map.[5] The detailed section below shows the cartographic symbolization; with, as Halley wrote, ". . . the sharp end of each little stroak pointing out that part of the Horizon, from whence the wind continually comes; and where there are Monsoons the rows of stroaks run alternately backwards and forwards, by which means they are thicker [denser] than elsewhere."

[5] Norman J. W. Thrower, "Edmond Halley as a Thematic Geo-Cartographer," *Annals of the Association of American Geographers*, 59 (December 1969), 652–676.

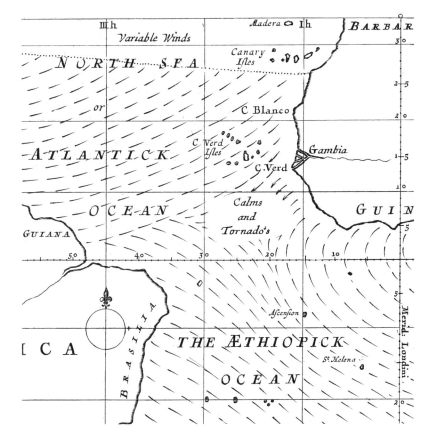

Edmond Halley, "An Historical Account of the Trade Winds, and Monsoons, Observable in the Seas Between and Near the Tropicks; With an Attempt to Assign the Phisical Cause of Said Winds," *Philosophical Transactions*, 183 (1686), 153–168.

An early and most worthy use of a map to chart patterns of
disease was the famous dot map of Dr. John Snow, who plotted
the location of deaths from cholera in central London for September 1854. Deaths were marked by dots and, in addition, the
area's eleven water pumps were located by crosses. Examining the
scatter over the surface of the map, Snow observed that cholera
occurred almost entirely among those who lived near (and drank
from) the Broad Street water pump. He had the handle of the
contaminated pump removed, ending the neighborhood epidemic
which had taken more than 500 lives.[6] The pump is located at the
center of the map, just to the right of the D in BROAD STREET. Of
course the link between the pump and the disease might have been
revealed by computation and analysis without graphics, with some
good luck and hard work. But, here at least, graphical analysis
testifies about the data far more efficiently than calculation.

[6] E. W. Gilbert, "Pioneer Maps of Health
and Disease in England," *Geographical
Journal*, 124 (1958), 172–183.

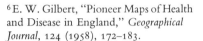

Charles Joseph Minard gave quantity as well as direction to the
data measures located on the world map in his portrayal of the
1864 exports of French wine:

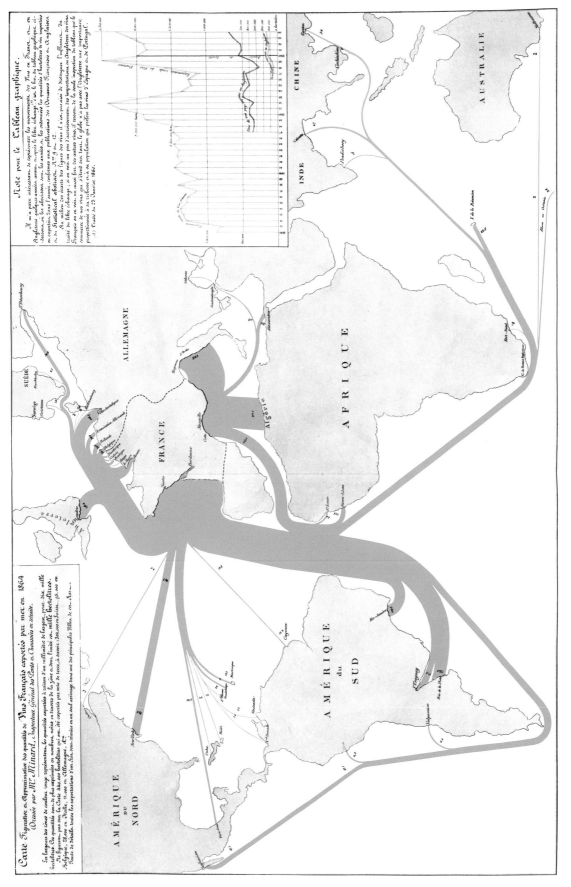

Charles Joseph Minard, *Tableaux Graphiques et Cartes Figuratives de M. Minard,*
1845–1869, a portfolio of his work held
by the Bibliothèque de l'École Nationale
des Ponts et Chaussées, Paris.

Computerized cartography and modern photographic techniques have increased the density of information some 5,000-fold in the best of current data maps compared to Halley's pioneering effort. This map shows the distribution of 1.3 million galaxies (including some overlaps) in the northern galactic hemisphere. The map divides the sky into 1,024×2,222 rectangles. The number of galaxies counted in each of the 2,275,328 rectangles is represented by ten gray tones; the darker the tone, the greater the number of galaxies counted. The north galactic pole is at the center. The sharp edge on the left results from the earth blocking the view from the observatory. In the area near the perimeter of the map, the view is obscured by the interstellar dust of the galaxy in which we live (the Milky Way) as the line of sight passes through the flattened disk of our galaxy. The curious texture of local clusters of galaxies seen in this truly new view of the universe was not anticipated by students of galaxies, who had, of course, microscopically examined millions of photographs of galaxies before seeing this macroscopic view. Although the clusters are clearly evident (and accounted for by a theory of galactic origins), the seemingly random filaments may be happenstance. The producers of the map note the "strong temptation to conclude that the galaxies are arranged in a remarkable filamentary pattern on scales of approximately 5° to 15°, but we caution that this visual impression may be misleading because the eye tends to pick out linear patterns even in random noise. Indeed, roughly similar patterns are seen on maps constructed from simulated catalogs where no linear structure has been built in. . . ."[7]

[7] Michael Seldner, B. H. Siebers, Edward J. Groth and P. James E. Peebles, "New Reduction of the Lick Catalog of Galaxies," *Astronomical Journal*, 82 (April 1977), 249–314.

The most extensive data maps, such as the cancer atlas and the count of the galaxies, place millions of bits of information on a single page before our eyes. No other method for the display of statistical information is so powerful.

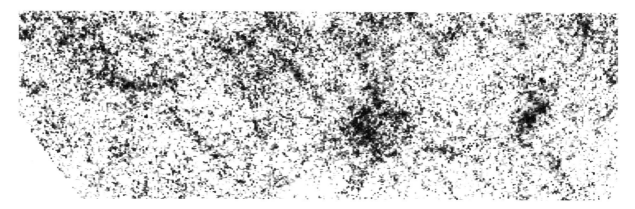

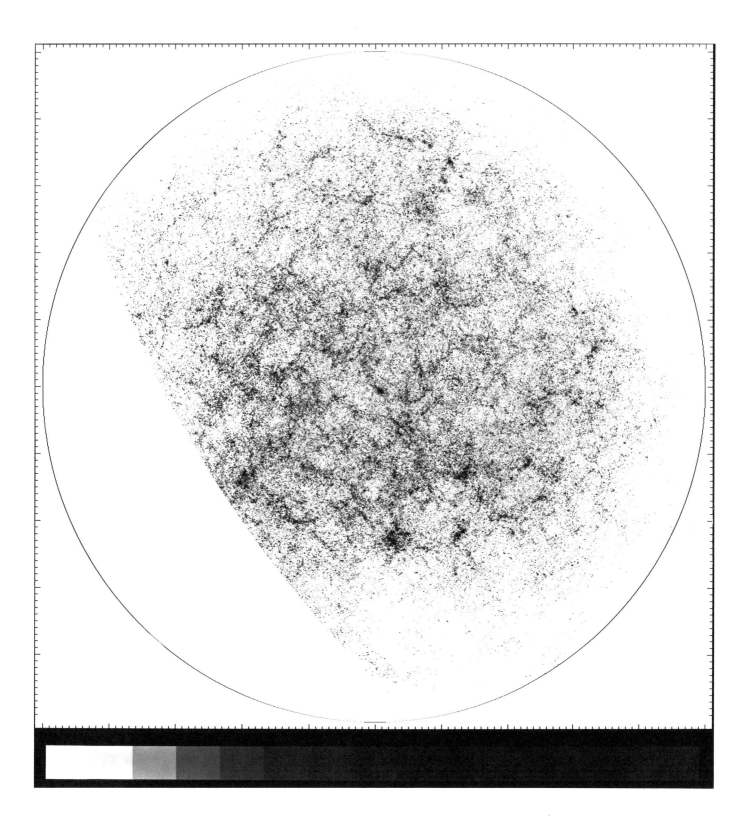

Time-Series

The time-series plot is the most frequently used form of graphic design.[8] With one dimension marching along to the regular rhythm of seconds, minutes, hours, days, weeks, months, years, centuries, or millennia, the natural ordering of the time scale gives this design a strength and efficiency of interpretation found in no other graphic arrangement.

This reputed tenth- (or possibly eleventh-) century illustration of the inclinations of the planetary orbits as a function of time, apparently part of a text for monastery schools, is the oldest known example of an attempt to show changing values graphically. It appears as a mysterious and isolated wonder in the history of data graphics, since the next extant graphic of a plotted time-series shows up some 800 years later. According to Funkhouser, the astronomical content is confused and there are difficulties in reconciling the graph and its accompanying text with the actual movements of the planets. Particularly disconcerting is the wavy path ascribed to the sun.[9] An erasure and correction of a curve occur near the middle of the graph.

[8] A random sample of 4,000 graphics drawn from 15 of the world's newspapers and magazines published from 1974 to 1980 found that more than 75 percent of all the graphics published were time-series. Chapter 3 reports more on this.

[9] H. Gray Funkhouser, "A Note on a Tenth Century Graph," *Osiris*, 1 (January 1936), 260–262.

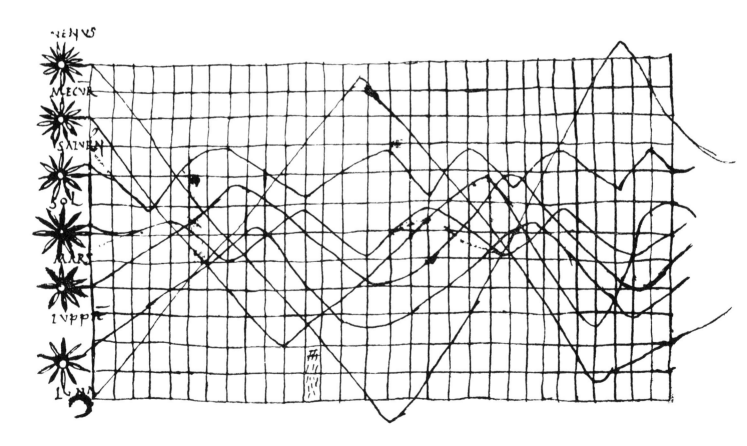

It was not until the late 1700s that time-series charts began to appear in scientific writings. This drawing of Johann Heinrich Lambert, one of a long series, shows the periodic variation in soil temperature in relation to the depth under the surface. The greater the depth, the greater the time-lag in temperature responsiveness. Modern graphic designs showing time-series periodicities differ little from those of Lambert, although the data bases are far larger.

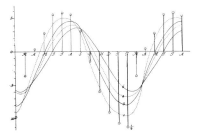

J. H. Lambert, *Pyrometrie* (Berlin, 1779).

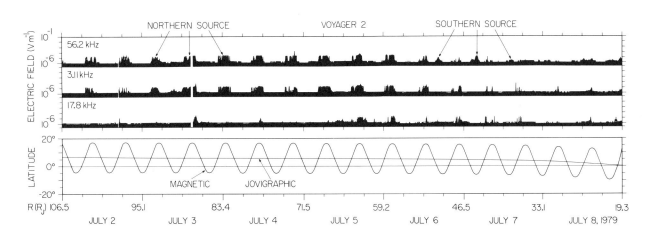

This plot of radio emissions from Jupiter is based on data collected by Voyager 2 in its pass close by the planet in July 1979. The radio intensity increases and decreases in a ten-hour cycle as Jupiter rotates. Maximum intensity occurs when the Jovian north magnetic pole is tipped toward the spacecraft, indicating a northern hemisphere source. A southern source was detected on July 7, as the spacecraft neared the equatorial plane. The horizontal scale shows the distance of the spacecraft from the planet measured in terms of Jupiter radii (R). Note the use of dual labels on the horizontal to indicate both the date and distance from Jupiter. The entire bottom panel also serves to label the horizontal scale, describing the changing orientation of the spacecraft relative to Jupiter as the planet is approached. The multiple time-series enforce not only comparisons within each series over time (as do all time-series plots) but also comparisons between the three different sampled radio bands shown. This richly multivariate display is based on 453,600 instrument samples of eight bits each. The resulting 3.6 million bits were reduced by peak and average processing to the 18,900 points actually plotted on the graphic.

D. A. Gurnett, W. S. Kurth, and F. L. Scarf, "Plasma Wave Observations Near Jupiter: Initial Results from Voyager 2," *Science* 206 (November 23, 1979), 987–991; and letter from Donald A. Gurnett to Edward R. Tufte, June 27, 1980.

Time-series displays are at their best for big data sets with real variability. Why waste the power of data graphics on simple linear changes,

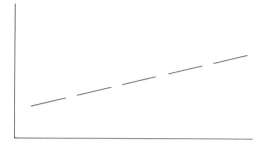

which can usually be better summarized in one or two numbers? Instead, graphics should be reserved for the richer, more complex, more difficult statistical material. This New York City weather summary for 1980 depicts 2,220 numbers. The daily high and low temperatures are shown in relation to the long-run average. The path of the normal temperatures also provides a forecast of expected change over the year; in the middle of February, for instance, New York City residents can look forward to warming at the rate of about 1.5 degrees per week all the way to July, the yearly peak. This distinguished graphic successfully organizes a large collection of numbers, makes comparisons between different parts of the data, and tells a story.

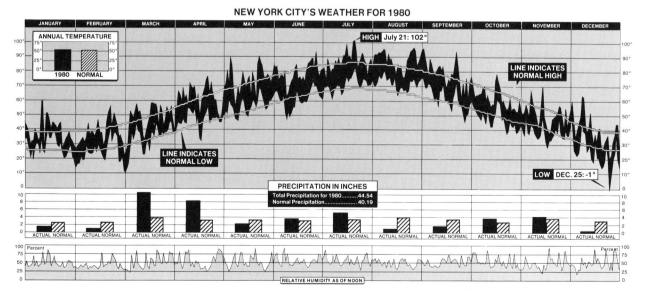

New York Times, January 11, 1981, p. 32.

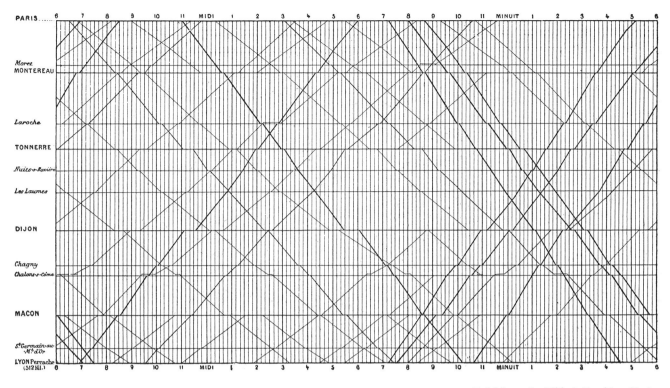

E. J. Marey, *La Méthode Graphique* (Paris, 1885), p. 20. The method is attributed to the French engineer, Ibry.

A design with similar strengths is Marey's graphical train schedule for Paris to Lyon in the 1880s. Arrivals and departures from a
station are located along the horizontal; length of stop at a station
is indicated by the length of the horizontal line. The stations are
separated in proportion to their actual distance apart. The slope
of the line reflects the speed of the train: the more nearly vertical
the line, the faster the train. The intersection of two lines locates
the time and place that trains going in opposite directions pass
each other.

In 1981 a new express train from Paris to Lyon cut the trip to
under three hours, compared to more than nine hours when Marey
published the graphical train schedule. The path of the modern
TGV (*train à grande vitesse*) is shown, overlaid on the schedule of
100 years before:

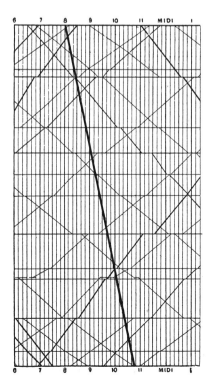

The two great inventors of modern graphical designs were J. H. Lambert (1728–1777), a Swiss-German scientist and mathematician, and William Playfair (1759–1823), a Scottish political economist.[10] The first known time-series using economic data was published in Playfair's remarkable book, *The Commercial and Political Atlas* (London, 1786). Note the graphical arithmetic, which shows the shifting balance of trade by the difference between the import and export time-series. Playfair contrasted his new graphical method with the tabular presentation of data:

> Information, that is imperfectly acquired, is generally as imperfectly retained; and a man who has carefully investigated a printed table, finds, when done, that he has only a very faint and partial idea of what he has read; and that like a figure imprinted on sand, is soon totally erased and defaced. The amount of mercantile transactions in money, and of profit or loss, are capable of being as easily represented in drawing, as any part of space, or as the face of a country; though, till now, it has not been attempted. Upon that principle these Charts were made; and, while they give a simple and distinct idea, they are as near perfect accuracy as is any way useful. On inspecting any one of these Charts attentively, a sufficiently distinct impression will be made, to remain unimpaired for a considerable time, and the idea which does remain will be simple and complete, at once including the duration and the amount. [pages 3–4]

For Playfair, graphics were preferable to tables because graphics showed the shape of the data in a comparative perspective. Time-

[10] Laura Tilling, "Early Experimental Graphs," *British Journal for the History of Science*, 8 (1975), 193–213.

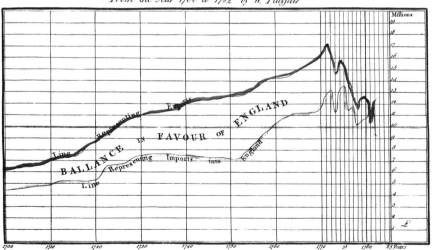

CHART of all the IMPORTS and EXPORTS to and from ENGLAND
From the Year 1700 to 1782 by W. Playfair

The Divisions at the Bottom, express YEARS, & those on the Right hand, MILLIONS of POUNDS

J. Ainslie Sculp.t   Publish'd as the Act directs, 20.th Aug.t 1785

series plots did this, and all but one of the 44 charts in the first edition of *The Commercial and Political Atlas* were time-series. That one exception is the first known bar chart, which Playfair invented because year-to-year data were missing and he needed a design to portray the one-year data that were available. Nonetheless he was skeptical about his innovation:

> This Chart is different from the others in principle, as it does not comprehend any portion of time, and it is much inferior in utility to those that do; for though it gives the extent of the different branches of trade, it does not compare the same branch of commerce with itself at different periods; nor does it imprint upon the mind that distinct idea, in doing which, the chief advantage of Charts consists: for as it wants the dimension that is formed by duration, there is no shape given to the quantities. [page 101]

He was right: small, noncomparative, highly labeled data sets usually belong in tables.

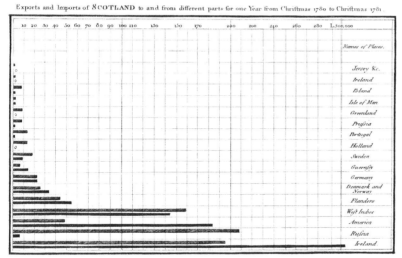

Exports and Imports of SCOTLAND to and from different parts for one Year from Christmas 1780 to Christmas 1781.

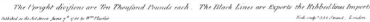

The Upright divisions are Ten Thousand Pounds each. The Black Lines are Exports the Ribbed lines Imports.

The chart does show, at any rate, the imports (cross-hatched lines) and exports (solid lines) to and from Scotland in 1781 for 17 countries, which are ordered by volume of trade. The horizontal scale is at the top, possibly to make it more convenient to see in plotting the points by hand. Zero values are nicely indicated both by the absence of a bar and by a "0." The horizontal scale mistakenly repeats "200." In nearly all his charts, Playfair placed the labels for the vertical scale on the right side of the page (suggesting that he plotted the data points using his left hand).

Playfair's last book addressed the question whether the price of wheat had increased relative to wages. In his *Letter on our agricultural distresses, their causes and remedies; accompanied with tables and copper-plate charts shewing and comparing the prices of wheat, bread and labour, from 1565 to 1821,* Playfair wrote:

> You have before you, my Lords and Gentlemen, a chart of the prices of wheat for 250 years, made from official returns; on the same plate I have traced a line representing, as nearly as I can, the wages of good mechanics, such as smiths, masons, and carpenters, in order to compare the proportion between them and the price of wheat at every different period. . . . the main fact deserving of consideration is, that never at any former period was wheat so cheap, in proportion to mechanical labour, as it is at the present time. . . . [pages 29–31]

Here Playfair plotted three parallel time-series: prices, wages, and the reigns of British kings and queens.

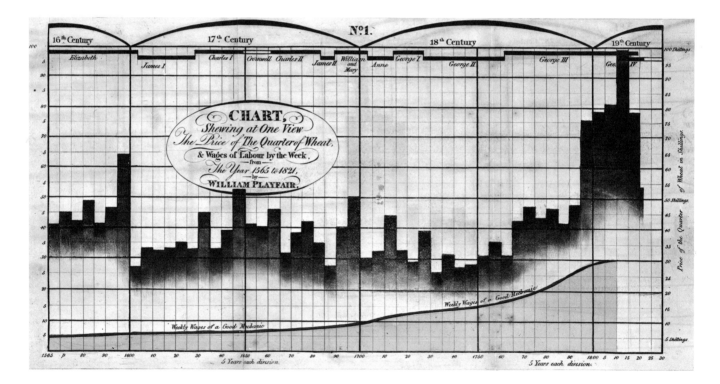

The history and genealogy of royalty was long a graphical favorite. This superb construction of E. J. Marey brings together several sets of facts about English rulers into a time-series that conveys a sense of the march of history. Marey (1830–1904) also pioneered the development of graphical methods in human and animal physiology, including studies of horses moving at different paces,

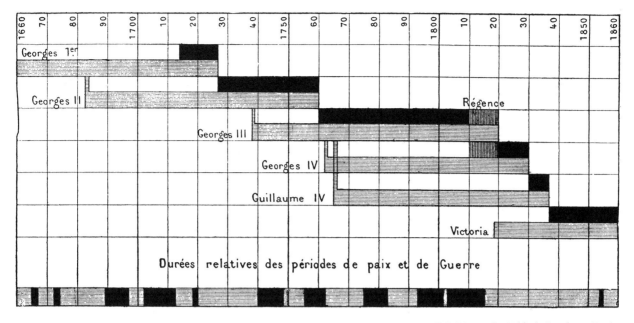

E. J. Marey, *La Méthode Graphique* (Paris, 1885), p. 6.

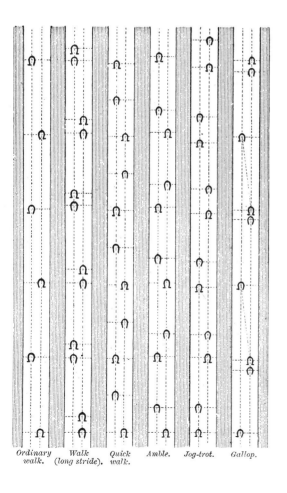

E. J. Marey, *Movement* (London, 1895). Beginning with the tracks of the horse, the time-series are from pages 191, 224, 222, 265, 60, and 61.

the movement of a starfish turning itself over (read images from the bottom upwards),

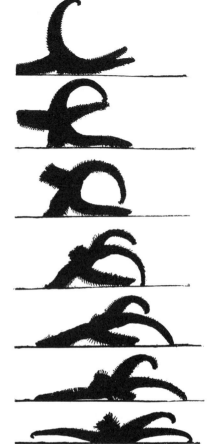

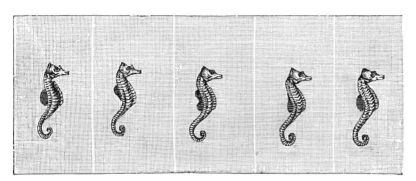

the undulations of the dorsal fin of a descending sea-horse,

as well as the advance of the gecko.

Marey's man in black velvet, photographed in stick-figure images, became the time-series forerunner of Marcel Duchamp's *Nude Descending a Staircase*.

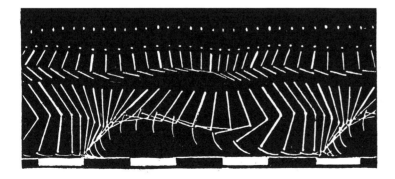

The problem with time-series is that the simple passage of time is not a good explanatory variable: descriptive chronology is not causal explanation. There are occasional exceptions, especially when there is a clear mechanism that drives the Y-variable. This time-series does testify about causality: the outgoing mail of the U.S. House of Representatives peaks every two years, just before the election day:

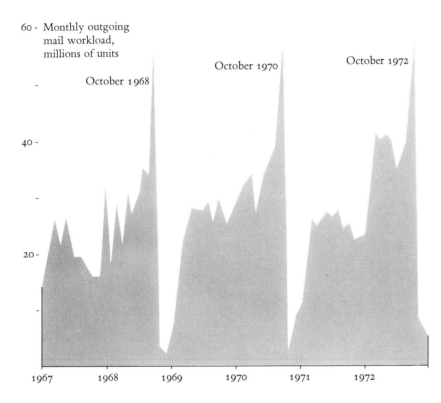

The graphic is worth at least 700 words, the number used in a news report describing how incumbent representatives exploit their free mailing privileges to advance their re-election campaigns:

FRANKED MAIL TIE TO VOTING SHOWN

Testimony Finds the Volume Rises Before Elections

WASHINGTON, June 1 (AP) —New court testimony and documents show that much of the mail Congress sends at taxpayer expense is tied directly to the re-election campaigns of Senate and House members. According to material filed in a lawsuit in Federal Court:

¶Senate Republicans put two direct-mail experts on the public payroll to advise them on how to use their free mailing privileges to get votes.

¶An election manual prepared for Senate Democrats refers to newsletters as a "free forum," and sets up a timetable for sending them as an integral part of a model re-election campaign.

¶Senator John G. Tower, Republican of Texas, mailed more than 800,000 special-interest letters at taxpayer expense as part of his 1972 re-election effort and received campaign volunteer offers and donations in response.

¶Senator Jacob K. Javits, Republican of New York, gave written approval in 1973 for a tax-paid mail program intended to better his image and pay off at the polls. He focused his mail on areas where he needed votes.

¶The volume of "official" Congressional mail rises in election years and peaks just before the general election.

None of this activity necessarily violates any law or regulation, since Congress has wide discretion in the use of tax-paid mail. Congress gave itself the right to send official mail at Government expense at the founding of the republic, and only Congress polices against abuses of the free mailings.

Complaints of political use of the franking privilege, called the franking privilege, are heard every election year. Recently, however, the volume and cost of franked mail has multiplied. A new Federal law will limit what out-of-office challengers can spend to unseat incumbents.

In 1972, Congress passed a law prohibiting mass franked mailings within 28 days before an election. The sponsor of that legislation, Representative Morris K. Udall, Democrat of Arizona, said in an interview that further changes were needed to curtail political abuse of the frank.

Mr. Udall urged a 60-day pre-election cutoff for mass mailings and said he favored closing a loophole that recently allowed defeated Representative Frank M. Clark, Democrat of Pennsylvania, to send a franked newsletter to his old constituents after he had left office. Mr. Clark is seeking to regain his old post.

Practice Documented

Seldom has the political use of franked mail been so well documented as in recent testimony and documents filed in a Federal Court by Common Cause, the lobby group, which is suing for an end to tax-financed mass mailings by Congress.

For example, Joyce P. Baker, a political mail specialist, said she wanted to set up direct-mail programs for Republican Senators using franked mail. "The purpose of such a program is to help an incumbent Senator get re-elected," she said.

She was put on the Senate payroll at $18,810 a year in 1973 and 1974 and testified that during that time she aided Republican Senators Robert J. Dole of Kansas, Peter H. Dominick of Colorado, Charles McC. Mathias Jr. of Maryland

Another political mail specialist, Lee W. MacGregor, wrote a proposal for the use of franked mail by his chief, Senator Javits, in 1973.

"The over-all objective of the franked mail program can be to get the recipient of the mail to identify positively with a particular stand you have taken or a bill you have introduced; the kind of identification that can be translated into a vote at the polls on election day," Mr. MacGregor said.

Mr. Javits was out of the country and could not be reached. His administrative assistant, Donald Kellerman, defended the use of franked mail.

"It is a standard device to let voters, not voters but citizens, know what the Senator is doing here in Washington," he said.

Senator Tower's use of franked mail in his 1972 campaign was documented by memorandums.

Tom Loeffler, a high-ranking campaign aide, wrote in a memorandum dated Oct. 27, 1972, that during the campaign Senator Tower had sent "31 special interest letters totaling approximately 803,333 franked mailings."

Mr. Tower was not available for comment. His administrative assistant, Elwin Skiles, said the Senator's use of franked mail in 1972 was within the law, and he defended the free-mailing privileges.

Postal Service figures show that in the 12 months before November, 1973, Congress sent 222.9 million franked pieces of mail. But in the next 12 months, covering the election season of 1974, Congress sent 350.6 million, a jump of 57 per cent about what's happening," Mr. Skiles said.

Time-series plots can be moved toward causal explanation by smuggling additional variables into the graphic design. For example, this decomposition of economic data, arraying 1,296 numbers, breaks out the top series into seasonal and trading-day fluctuations (which dominate short-term changes) to reveal the long-run trend adjusted for inflation. (Note a significant defect in the design, however: the vertical grid conceals the height of the December peaks.) The next step would be to bring in additional variables to explain the transformed and improved series at the bottom.[11]

[11] See William S. Cleveland and Irma J. Terpenning, "Graphical Methods for Seasonal Adjustment," *Journal of the American Statistical Association* 77 (March 1982), 52–62.

Julius Shiskin, "Measuring Current Economic Fluctuations," *Statistical Reporter* (July 1973), p. 3.

Systematic and Irregular Components of Total Retail Sales, United States

Finally, a vivid design (with appropriate data) is the before-after time-series:

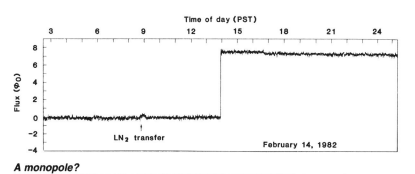

A monopole?

Cabrera's candidate monopole signal looms over a disturbance caused by a liquid nitrogen transfer earlier in the day. The jump in magnetic flux through the superconducting detector loop (or equivalently, the jump in the loop's supercurrent) is just the right magnitude to be a monopole. Moreover, the current remained stable for many hours afterward.

M. Mitchell Waldrop, "In Search of the Magnetic Monopole," *Science* (June 4, 1982), p. 1087.

And before and after the collapse of a bridge on the Rhône in 1840:

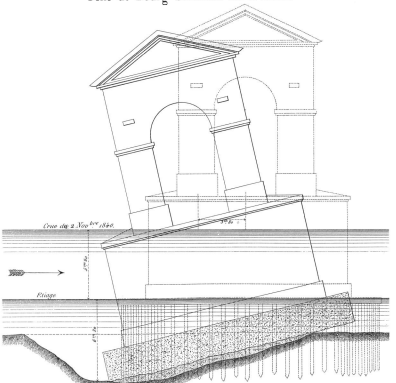

Pont de **Bourg-St Andéol** sur le **Rhône**.

Charles Joseph Minard, "De la Chute des Ponts dans les grandes Crues," (October 24, 1856), Figure 3, in Minard, *Collection de ses brochures* (Paris, 1821–1869), held by the Bibliothèque de l'École Nationale des Ponts et Chaussées, Paris.

Narrative Graphics of Space and Time

An especially effective device for enhancing the explanatory power of time-series displays is to add spatial dimensions to the design of the graphic, so that the data are moving over space (in two or three dimensions) as well as over time. Three excellent space-time-story graphics illustrate here how multivariate complexity can be subtly integrated into graphical architecture, integrated so gently and unobtrusively that viewers are hardly aware that they are looking into a world of four or five dimensions. Occasionally graphics are belligerently multivariate, advertising the technique rather than the data. But not these three.

The first is the classic of Charles Joseph Minard (1781–1870), the French engineer, which shows the terrible fate of Napoleon's army in Russia. Described by E. J. Marey as seeming to defy the pen of the historian by its brutal eloquence,[12] this combination of data map and time-series, drawn in 1861, portrays the devastating losses suffered in Napoleon's Russian campaign of 1812. Beginning at the left on the Polish-Russian border near the Niemen River, the thick band shows the size of the army (422,000 men) as it invaded Russia in June 1812. The width of the band indicates the size of the army at each place on the map. In September, the army reached Moscow, which was by then sacked and deserted, with 100,000 men. The path of Napoleon's retreat from Moscow is depicted by the darker, lower band, which is linked to a temperature scale and dates at the bottom of the chart. It was a bitterly cold winter, and many froze on the march out of Russia. As the graphic shows, the crossing of the Berezina River was a disaster, and the army finally struggled back into Poland with only 10,000 men remaining. Also shown are the movements of auxiliary troops, as they sought to protect the rear and the flank of the advancing army. Minard's graphic tells a rich, coherent story with its multivariate data, far more enlightening than just a single number bouncing along over time. *Six* variables are plotted: the size of the army, its location on a two-dimensional surface, direction of the army's movement, and temperature on various dates during the retreat from Moscow.

It may well be the best statistical graphic ever drawn.

[12] E. J. Marey, *La Méthode Graphique* (Paris, 1885), p. 73. For more on Minard, see Arthur H. Robinson, "The Thematic Maps of Charles Joseph Minard," *Imago Mundi*, 21 (1967), 95–108.

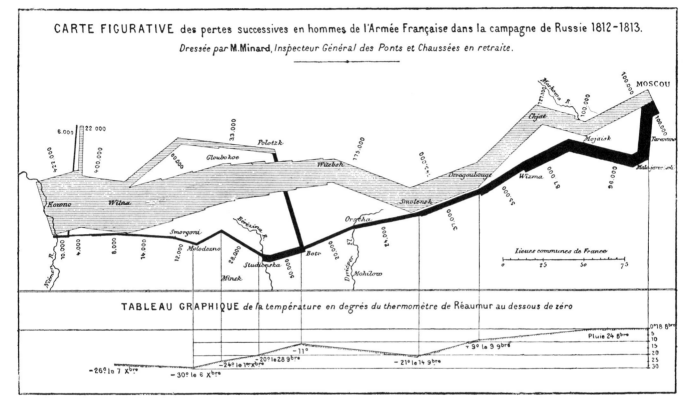

CARTE FIGURATIVE des pertes successives en hommes de l'Armée Française dans la campagne de Russie 1812-1813.

Dressée par M. Minard, Inspecteur Général des Ponts et Chaussées en retraite.

TABLEAU GRAPHIQUE de la température en degrés du thermomètre de Réaumur au dessous de zéro

X^{bre} = December   9bre = November   8bre = October

The next time-space graphic, drawn by a computer, displays the levels of three air pollutants located over a two-dimensional surface (six counties in southern California) at four times during the day. Nitrogen oxides (top row) are emitted by power plants, refineries, and vehicles. Refineries along the coast and Kaiser Steel's Fontana plant produce the post-midnight peaks shown in the first panel; traffic and power plants (with their heavy daytime demand) send levels up during the day. Carbon monoxide (second row) is low after midnight except out at the steel plant; morning traffic then begins to generate each day's ocean of carbon monoxide, with the greatest concentration at the convergence of five freeways in downtown Los Angeles. Reactive hydrocarbons (third row), like nitrogen oxides, come from refineries after midnight and then increase with traffic during the day. Each of the 12 time-space-pollutant slices summarizes pollutants for 2,400 spatial locations (2,400 squares five kilometers on a side). Thus 28,800 pollutant readings are shown, except for those masked by peaks.

The air pollution display is a *small multiple*. The same graphical design structure is repeated for each of the twelve slices or multiples. Small multiples are economical: once viewers understand the design of one slice, they have immediate access to the data in all the other slices. Thus, as the eye moves from one slice to the next, the constancy of the design allows the viewer to focus on changes in the data rather than on changes in graphical design.

Los Angeles Times, July 22, 1979; based on work of Gregory J. McRae, California Institute of Technology.

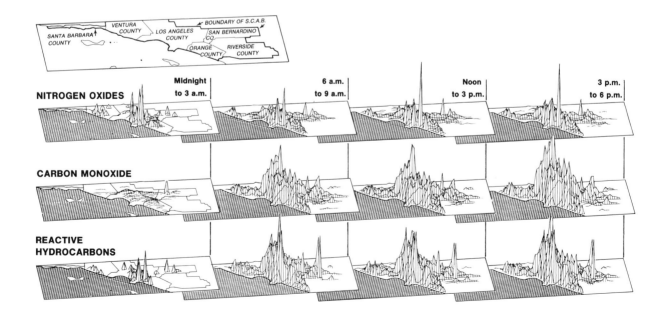

Our third example of a space-time-story graphic ingeniously mixes space and time on the horizontal axis. This design moves well beyond the conventional time-series because of its clever plotting field, with location relative to the ground surface on the vertical axis and time/space on the horizontal. The life cycle of the Japanese beetle is shown.

L. Hugh Newman, *Man and Insects* (London, 1965), pp. 104–105.

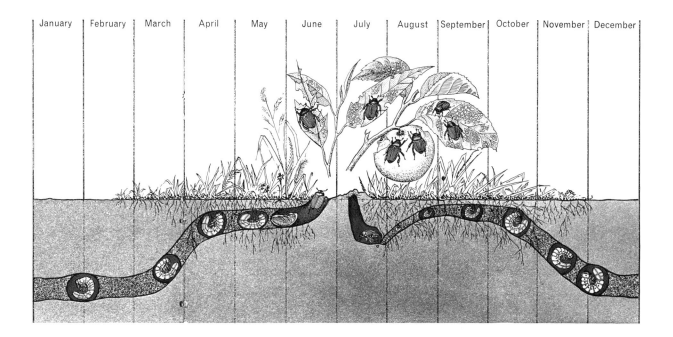

| January | February | March | April | May | June | July | August | September | October | November | December |

More Abstract Designs: Relational Graphics

The invention of data graphics required replacing the latitude-longitude coordinates of the map with more abstract measures not based on geographical analogy. Moving from maps to statistical graphics was a big step, and thousands of years passed before this step was taken by Lambert, Playfair, and others in the eighteenth century. Even so, analogies to the physical world served as the conceptual basis for early time-series. Playfair repeatedly compared his charts to maps and, in the preface to the first edition of *The Commercial and Political Atlas*, argued that his charts corresponded to a physical realization of the data:

> Suppose the money we pay in any one year for the expence of the Navy were in guineas, and that these guineas were laid down upon a large table in a straight line, and touching each other, and those paid next year were laid down in another

straight line, and the same continued for a number of years: these lines would be of different lengths, as there were fewer or more guineas; and they would make a shape, the dimensions of which would agree exactly with the amount of the sums; and the value of a guinea would be represented by the part of space which it covered. The Charts are exactly this upon a small scale, and one division represents the breadth or value of ten thousand or an hundred thousand guineas as marked, with the same exactness that a square inch upon a map may represent a square mile of country. And they, therefore, are a representation of the real money laid down in different lines, as it was originally paid away. [pages iii–iv]

Fifteen years later in *The Statistical Breviary*, his most theoretical book about graphics, Playfair broke free of analogies to the physical world and drew graphics as designs-in-themselves.

One of four plates in *The Statistical Breviary*, this graphic is distinguished by its multivariate data, the use of area to depict quantity, and the pie chart—in apparently the first application of these devices. The circle represents the area of each country; the line on the left, the population in millions read on the vertical scales; the line on the right, the revenue (taxes) collected in millions of pounds sterling read also on the vertical scale; and the "dotted lines drawn between the population and revenue, are merely intended to connect together the lines belonging to the same country. The ascent

THE
STATISTICAL BREVIARY;
SHEWING,
ON A PRINCIPLE ENTIRELY NEW,
THE RESOURCES
OF EVERY
STATE AND KINGDOM IN EUROPE;
ILLUSTRATED WITH
STAINED COPPER-PLATE CHARTS,
REPRESENTING THE
PHYSICAL POWERS OF EACH DISTINCT NATION
WITH EASE AND PERSPICUITY.
By *WILLIAM PLAYFAIR.*
TO WHICH IS ADDED,
A SIMILAR EXHIBITION OF THE RULING POWERS
OF HINDOOSTAN.
LONDON:
Printed by T. Bensley, Bolt Court, Fleet Street,
For J. Wallis, 46, Paternoster Row; Carpenter and Co. Bond
Street; Egerton, Whitehall; Vernor and Hood, Poultry; Black
and Parry, Leadenhall Street; and Tibbet and Didier, St. James's
Street.
1801.

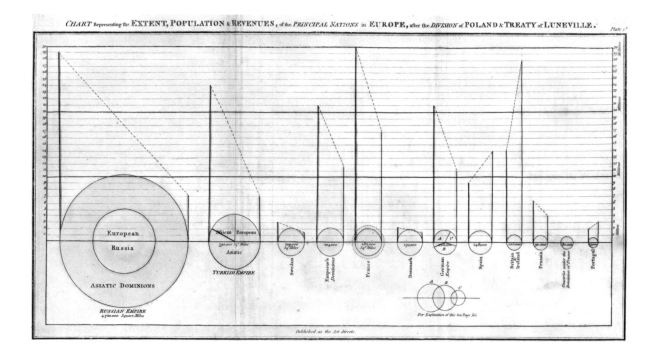

CHART Representing the EXTENT, POPULATION & REVENUES, of the PRINCIPAL NATIONS in EUROPE, after the DIVISION of POLAND & TREATY of LUNEVILLE.

of those lines being from right to left, or from left to right, shews whether in proportion to its population the country is burdened with heavy taxes or otherwise" (pages 13–14). The slope of the dotted line is uninformative, since it is dependent on the diameter of the circle as well as the height of the two verticals. However, the sign of the slope does make sense, taking Playfair to his familiar point about what he regarded as excessive taxation in Britain (fourth circle from the right, with the slope running opposite to most countries). Playfair was enthusiastic about the multivariate arrangement because it fostered comparisons:

> The author of this work applied the use of lines to matters of commerce and finance about fifteen years ago, with great success. His mode was generally approved of as not only facilitating, but rendering those studies more clear, and retained more easily by the memory. The present charts are in like manner intended to aid statistical studies, by shewing to the eye the sizes of different countries represented by similar forms, for where forms are not similar, the eye cannot compare them easily nor accurately. From this circumstance it happens, that we have a more accurate idea of the sizes of the planets, which are spheres, than of the nations of Europe which we see on the maps, all of which are irregular forms in themselves as well as unlike to each other. Size, Population, and Revenue, are the three principal objects of attention upon the general scale of statistical studies, whether we are actuated by curiosity or interest; I have therefore represented these three objects in one view. . . . [page 15]

But here Playfair had a forerunner—and one who thought more clearly about the abstract problems of graphical design than did Playfair, who lacked mathematical skills. A most remarkable and explicit early theoretical statement advancing the general (non-analogical) relational graphic was made by J. H. Lambert in 1765, 35 years before *The Statistical Breviary*:

> We have in general two variable quantities, x, y, which will be collated with one another by observation, so that we can determine for each value of x, which may be considered as an abscissa, the corresponding ordinate y. Were the experiments or observations completely accurate, these ordinates would give a number of points through which a straight or curved line should be drawn. But as this is not so, the line deviates to a greater or lesser extent from the observational points. It must therefore be drawn in such a way that it comes as near as possible to its true position and goes, as it were, through the middle of the given points.[13]

[13] Johann Heinrich Lambert, *Beyträge zum Gebrauche der Mathematik und deren Anwendung* (Berlin, 1765), as quoted in Laura Tilling, "Early Experimental Graphs," *British Journal for the History of Science*, 8 (1975), 204–205.

Lambert drew a graphical derivation of the evaporation rate of water as a function of temperature, according to Tilling. The analysis begins with two time-series: DEF, showing the decreasing height of water in a capillary tube as a function of time, and ABC, the temperature. The slope of the curve DEF is then taken (note the tangent DEG) at a number of places, yielding the rate of evaporation:

J. H. Lambert, "Essai d'hygrométrie ou sur la mesure de l'humidité," *Mémoires de l'Académie Royale des Sciences et Belles-Lettres . . . 1769* (Berlin, 1771), plate 1, facing p. 126; from Tilling's article.

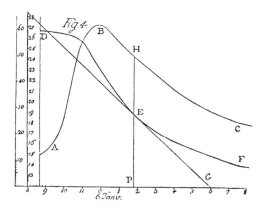

To complete the graphical calculus, the measured rate is plotted against the corresponding temperature in this relational graphic:

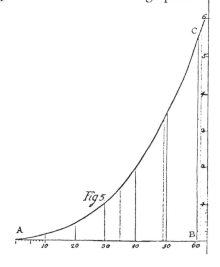

Thus, by the early 1800s, graphical design was at last no longer dependent on direct analogy to the physical world—thanks to the work of Lambert and Playfair. This meant, quite simply but quite profoundly, that any variable quantity could be placed in relationship to any other variable quantity, measured for the same units of observation. Data graphics, because they were relational and not tied to geographic or time coordinates, became relevant to all quantitative inquiry. Indeed, in modern scientific literature, about 40 percent of published graphics have a relational form, with two or more variables (none of which are latitude, longitude, or time). This is no accident, since the relational graphic—in its barest form, the scatterplot and its variants—is the greatest of all graphical designs. It links at least two variables, encouraging and even imploring the viewer to assess the possible causal relationship between the plotted variables. It confronts causal theories that X causes Y with empirical evidence as to the actual relationship between X and Y, as in the case of the relationship between lung cancer and smoking:

CRUDE MALE DEATH RATE FOR LUNG CANCER IN 1950 AND PER CAPITA CONSUMPTION OF CIGARETTES IN 1930 IN VARIOUS COUNTRIES.

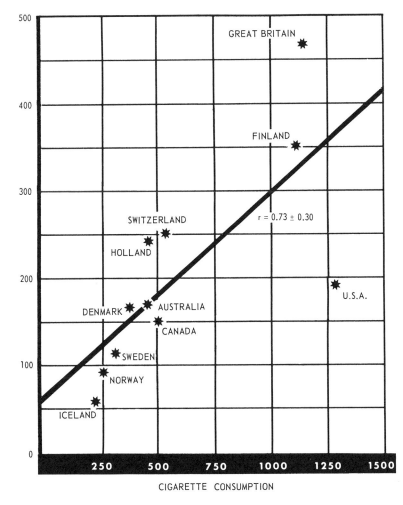

Report of the Advisory Committee to the Surgeon General, *Smoking and Health* (Washington, D.C., 1964), p. 176; based on R. Doll, "Etiology of Lung Cancer," *Advances in Cancer Research*, 3 (1955), 1–50.

These small-multiple relational graphs show unemployment and inflation over time in "Phillips curve" plots for nine countries, demonstrating the collapse of what was once thought to be an inverse relationship between the variables.

Paul McCracken, et al., *Towards Full Employment and Price Stability* (Paris, 1977), p. 106.

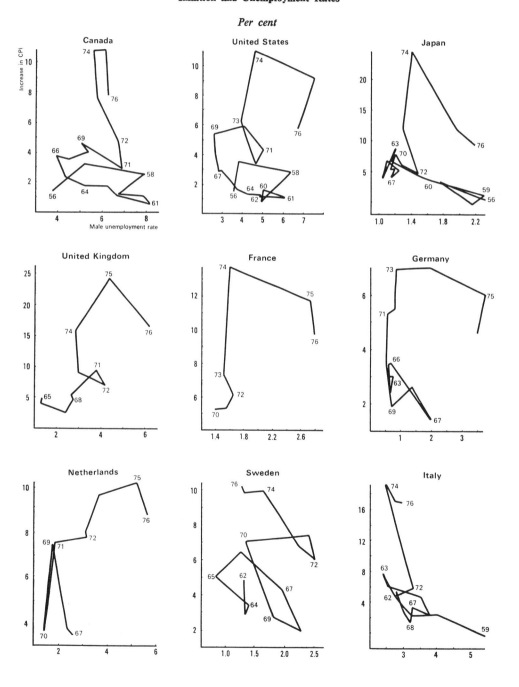

Inflation and Unemployment Rates

Per cent

Theory and measured observations diverge in the physical sciences, also. Here the relationship between temperature and the thermal conductivity of copper is assessed in a series of measurements from different laboratories. The connected points are from a single publication, cited by an identification number. The very different answers reported in the published literature result mainly from impurities in the samples of copper. Note how effectively the graphic organizes a vast amount of data, recording findings of hundreds of studies on a single page and, at the same time, enforcing comparisons of the varying results.

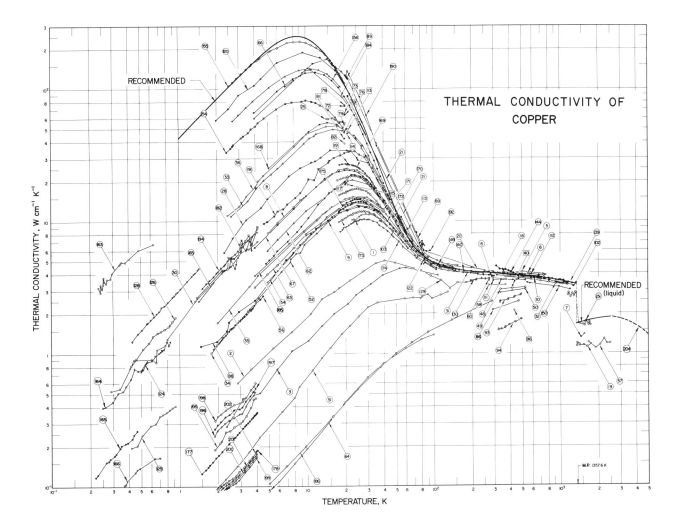

C. Y. Ho, R. W. Powell, and P. E. Liley, *Thermal Conductivity of the Elements: A Comprehensive Review*, supplement no. 1, *Journal of Physical and Chemical Reference Data*, 3 (1974), 1–244.

Finally, two relational designs of a different sort—wherein the data points are themselves data. Here the effect of two variables interacting is portrayed by the faces on the plotting field:

E. C. Zeeman, "Catastrophe Theory," *Scientific American*, 234 (April 1976), 67.

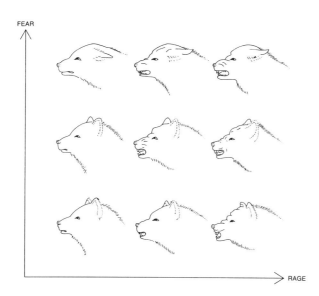

And similarly, the varying sizes of white pine seedlings after growing for one season in sand containing different amounts of calcium, in parts per million in nutrient-sand cultures:

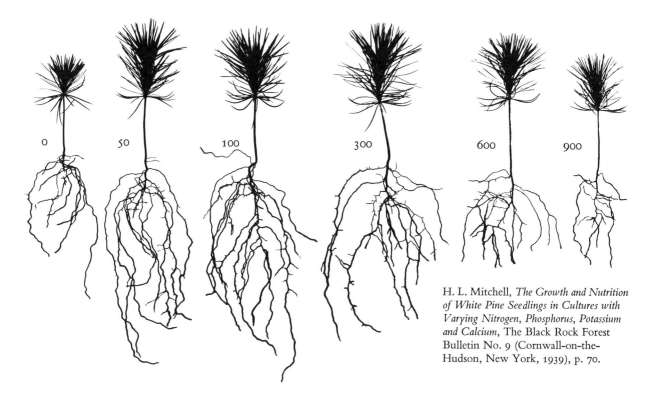

H. L. Mitchell, *The Growth and Nutrition of White Pine Seedlings in Cultures with Varying Nitrogen, Phosphorus, Potassium and Calcium*, The Black Rock Forest Bulletin No. 9 (Cornwall-on-the-Hudson, New York, 1939), p. 70.

Principles of Graphical Excellence

Graphical excellence is the well-designed presentation of interesting data—a matter of *substance*, of *statistics*, and of *design*.

Graphical excellence consists of complex ideas communicated with clarity, precision, and efficiency.

Graphical excellence is that which gives to the viewer the greatest number of ideas in the shortest time with the least ink in the smallest space.

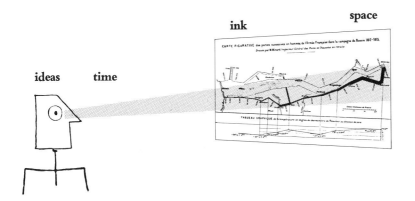

Graphical excellence is nearly always multivariate.

And graphical excellence requires telling the truth about the data.

As to the propriety and justness of representing sums of money, and time, by parts of space, tho' very readily agreed to by most men, yet a few seem to apprehend that there may possibly be some deception in it, of which they are not aware. . . .

William Playfair, *The Commercial and Political Atlas* (London, 1786)

People said: "With the chart on the wall, with the figures published, let's emulate and rouse our enthusiasm in production."

State Statistical Bureau of the People's Republic of China,
Statistical Work in the New China (Beijing, 1979)

Get it right or let it alone.
The conclusion you jump to may be your own.

James Thurber, *Further Fables for Our Time* (New York, 1956)

2 *Graphical Integrity*

For many people the first word that comes to mind when they think about statistical charts is "lie." No doubt some graphics do distort the underlying data, making it hard for the viewer to learn the truth. But data graphics are no different from words in this regard, for any means of communication can be used to deceive. There is no reason to believe that graphics are especially vulnerable to exploitation by liars; in fact, most of us have pretty good graphical lie detectors that help us see right through frauds.

Much of twentieth-century thinking about statistical graphics has been preoccupied with the question of how some amateurish chart might fool a naive viewer. Other important issues, such as the use of graphics for serious data analysis, were largely ignored. At the core of the preoccupation with deceptive graphics was the assumption that data graphics were mainly devices for showing the obvious to the ignorant. It is hard to imagine any doctrine more likely to stifle intellectual progress in a field. The assumption led down two fruitless paths in the graphically barren years from 1930 to 1970: First, that graphics had to be "alive," "communicatively dynamic," overdecorated and exaggerated (otherwise all the dullards in the audience would fall asleep in the face of those boring statistics). Second, that the main task of graphical analysis was to detect and denounce deception (the dullards could not protect themselves).

Then, in the late 1960s, John Tukey made statistical graphics respectable, putting an end to the view that graphics were only for decorating a few numbers. For here was a world-class data analyst spinning off half a dozen new designs and, more importantly, using them effectively to explore complex data.[1] Not a word about deception; no tortured attempts to construct more "graphical standards" in a hopeless effort to end all distortions. Instead, graphics were used as instruments for reasoning about quantitative information. With this good example, graphical work has come to flourish.

Of course false graphics are still with us. Deception must always be confronted and demolished, even if lie detection is no longer at the forefront of research. Graphical excellence begins with telling the truth about the data.

[1]John W. Tukey and Martin B. Wilk, "Data Analysis and Statistics: Techniques and Approaches," in Edward R. Tufte, ed., *The Quantitative Analysis of Social Problems* (Reading, Mass., 1970), 370–390; and John W. Tukey, "Some Graphic and Semigraphic Displays," in T. A. Bancroft, ed., *Statistical Papers in Honor of George W. Snedecor* (Ames, Iowa, 1972), 293–316.

Here are several graphics that fail to tell the truth. First, the case of the disappearing baseline in the annual report of a company that would just as soon forget about 1970. A careful look at the middle panel reveals a negative income in 1970, which is disguised by having the bars begin at the bottom at approximately minus $4,200,000:

Day Mines, Inc., *1974 Annual Report*, p. 1.

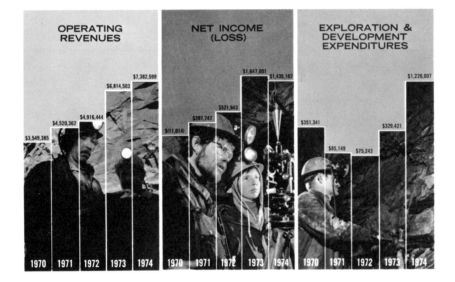

This pseudo-decline was created by comparing six months' worth of payments in 1978 to a full year's worth in 1976 and 1977, with the lie repeated four times over.

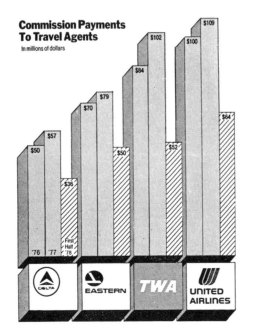

New York Times, August 8, 1978, p. D-1.

And sometimes the fact that numbers have a magnitude as well as an order is simply forgotten:

Comparative Annual Cost per Capita for care of Insane in Pittsburgh City Homes and Pennsylvania State Hospitals.

$147 South Mountain $172 Pittsburgh $198 Harrisburg $213 Norristown $214 Warren

Pittsburgh Civic Commission, *Report on Expenditures of the Department of Charities* (Pittsburgh, 1911), p. 7.

What is Distortion in a Data Graphic?

A graphic does not distort if the visual representation of the data is consistent with the numerical representation. What then is the "visual representation" of the data? As physically measured on the surface of the graphic? Or the *perceived* visual effect? How do we know that the visual image represents the underlying numbers?

One way to try to answer these questions is to conduct experiments on the visual perception of graphics—having people look at lines of varying length, circles of different areas, and then recording their assessments of the numerical quantities.

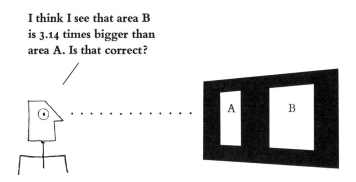

I think I see that area B is 3.14 times bigger than area A. Is that correct?

Such experiments have discovered very approximate power laws relating the numerical measure to the reported perceived measure. For example, the perceived area of a circle probably grows somewhat more slowly than the actual (physical, measured) area: the reported perceived area = (actual area)^x, where x = .8±.3, a discouraging result. Different people see the same areas somewhat differently; perceptions change with experience; and perceptions are context-dependent.[2] Particularly disheartening is the securely established finding that the reported perception of something as clear and simple as line length depends on the context and what other people have already said about the lines.[3]

Misperception and miscommunication are certainly not special to statistical graphics,

but what is a poor designer to do? A different graphic for each perceiver in each context? Or designs that correct for the visual transformations of the average perceiver participating in the average psychological experiment?

One satisfactory answer to these questions is to use a table to show the numbers. Tables usually outperform graphics in reporting on small data sets of 20 numbers or less. The special power of graphics comes in the display of large data sets.

At any rate, given the perceptual difficulties, the best we can hope for is some uniformity in graphics (if not in the perceivers) and some assurance that perceivers have a fair chance of getting the numbers right. Two principles lead toward these goals and, in consequence, enhance graphical integrity:

> The representation of numbers, as physically measured on the surface of the graphic itself, should be directly proportional to the numerical quantities represented.

> Clear, detailed, and thorough labeling should be used to defeat graphical distortion and ambiguity. Write out explanations of the data on the graphic itself. Label important events in the data.

[2] The extensive literature is summarized in Michael Macdonald-Ross, "How Numbers Are Shown: A Review of Research on the Presentation of Quantitative Data in Texts," *Audio-Visual Communication Review*, 25 (1977), 359–409. In particular, H. J. Meihoefer finds great variability among perceivers; see "The Utility of the Circle as an Effective Cartographic Symbol," *Canadian Cartographer*, 6 (1969), 105–117; and "The Visual Perception of the Circle in Thematic Maps: Experimental Results," ibid., 10 (1973), 63–84.

[3] S. E. Asch, "Studies of Independence and Submission to Group Pressure. A Minority of One Against a Unanimous Majority," *Psychological Monographs* (1956), 70.

Drawing by CEM; copyright 1961, *The New Yorker*.

Violations of the first principle constitute one form of graphic misrepresentation, measured by the

$$\text{Lie Factor} = \frac{\text{size of effect shown in graphic}}{\text{size of effect in data}}$$

If the Lie Factor is equal to one, then the graphic might be doing a reasonable job of accurately representing the underlying numbers. Lie Factors greater than 1.05 or less than .95 indicate substantial distortion, far beyond minor inaccuracies in plotting. The logarithm of the Lie Factor can be taken in order to compare overstating (log LF > 0) with understating (log LF < 0) errors. In practice almost all distortions involve overstating, and Lie Factors of two to five are not uncommon.

Here is an extreme example. A newspaper reported that the U.S. Congress and the Department of Transportation had set a series of fuel economy standards to be met by automobile manufacturers, beginning with 18 miles per gallon in 1978 and moving in steps up to 27.5 by 1985, an increase of 53 percent:

$$\frac{27.5 - 18.0}{18.0} \times 100 = 53\%$$

These standards and the dates for their attainment were shown:

This line, representing 18 miles per gallon in 1978, is 0.6 inches long.

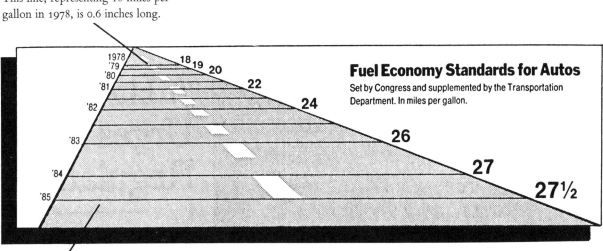

This line, representing 27.5 miles per gallon in 1985, is 5.3 inches long.

New York Times, August 9, 1978, p. D-2.

The magnitude of the change from 1978 to 1985 is shown in the graph by the relative lengths of the two lines:

$$\frac{5.3 - 0.6}{0.6} \times 100 = 783\%$$

Thus the numerical change of 53 percent is presented by some lines that changed 783 percent, yielding

$$\text{Lie Factor} = \frac{783}{53} = 14.8$$

which is too big.

The display also has several peculiarities of perspective:

· On most roads the future is in front of us, toward the horizon, and the present is at our feet. This display reverses the convention so as to exaggerate the severity of the mileage standards.

· Oddly enough, the dates on the left remain a constant size on the page even as they move along with the road toward the horizon.

· The numbers on the right, as well as the width of the road itself, are shrinking because of two simultaneous effects: the change in the values portrayed and the change due to perspective. Viewers have no chance of separating the two.

It is easy enough to decorate these data without lying:

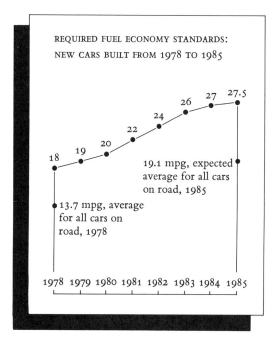

The non-lying version, in addition, puts the data in a context by comparing the new car standards with the mileage achieved by the mix of cars actually on the road. Also revealed is a side of the data disguised and mispresented in the original display: the fuel economy standards require gradual improvement at start-up, followed by a doubled rate from 1980 to 1983, and flattening out after that.

Sometimes decoration can help editorialize about the substance of the graphic. But it is wrong to distort the data measures—the ink locating values of numbers—in order to make an editorial comment or fit a decorative scheme. It is also a sure sign of the Graphical Hack at work. Here are many decorations but no lies:

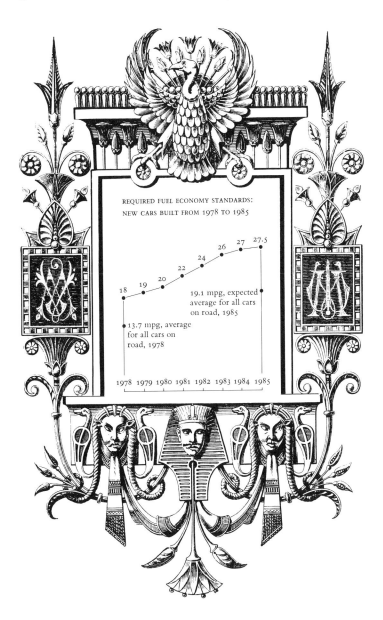

REQUIRED FUEL ECONOMY STANDARDS:
NEW CARS BUILT FROM 1978 TO 1985

18 19 20 22 24 26 27 27.5

19.1 mpg, expected
average for all cars
on road, 1985

13.7 mpg, average
for all cars on
road, 1978

1978 1979 1980 1981 1982 1983 1984 1985

Design and Data Variation

Each part of a graphic generates visual expectations about its other parts and, in the economy of graphical perception, these expectations often determine what the eye sees. Deception results from the incorrect extrapolation of visual expectations generated at one place on the graphic to other places.

A scale moving in regular intervals, for example, is expected to continue its march to the very end in a consistent fashion, without the muddling or trickery of non-uniform changes. Here an irregular scale is used to concoct a pseudo-decline. The first seven increments on the horizontal scale are ten years long, masking the rightmost interval of four years. Consequently the conspicuous feature of the graphic is the apparent fall of curves at the right, particularly the decline in prizes won by people from the United States (the heavy, dark line) in the most recent period. This effect results solely from design variation. It is a big lie, since in reality (and even in extrapolation, scaling up each end-point by 2.5 to take the four years' worth of data up to a comparable decade), the U.S. curve turned sharply upward in the post-1970 interval. A correction, with the actual data for 1971–80, is at the right:

National Science Foundation, *Science Indicators, 1974* (Washington, D.C., 1976), p. 15.

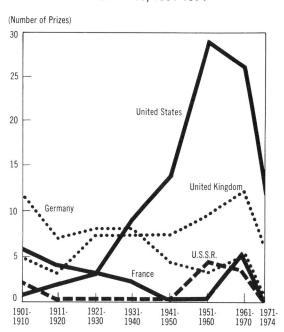

Nobel Prizes Awarded in Science, for Selected Countries, 1901-1974

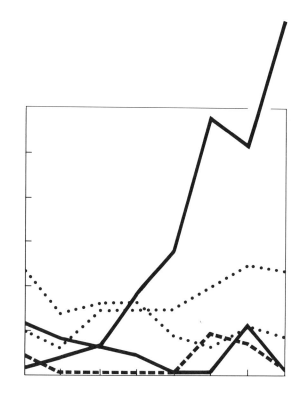

The confounding of *design variation* with *data variation* over the surface of a graphic leads to ambiguity and deception, for the eye may mix up changes in the design with changes in the data. A steady canvas makes for a clearer picture. The principle is, then:

Show data variation, not design variation.

Design variation corrupts this display:

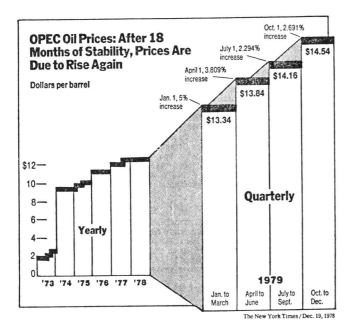

New York Times, December 19, 1978, p. D-7.

Five different vertical scales show the price:

During this time	one vertical inch equals
1973–1978	$8.00
January–March 1979	$4.73
April–June 1979	$4.37
July–September 1979	$4.16
October–December 1979	$3.92

And two different horizontal scales show the passage of time:

During this time	one horizontal inch equals
1973–1978	3.8 years
1979	0.57 years

As the two scales shift simultaneously, the distortion takes on multiplicative force. On the left of the graph, a price of $10 for one year is represented by 0.31 square inches; on the right side, by 4.69 square inches. Thus exactly the same quantity is 4.69/0.31 =15.1 times larger depending upon where it happens to fall on the surface of the graphic. *That* is design variation.

Design variation infected similar graphics in other publications. Here an increase of 454 percent is depicted as an increase of 4,280 percent, for a Lie Factor of 9.4:

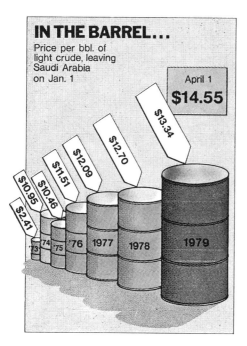

Time, April 9, 1979, p. 57.

And an increase of 708 percent is shown as 6,700 percent, for a Lie Factor of 9.5:

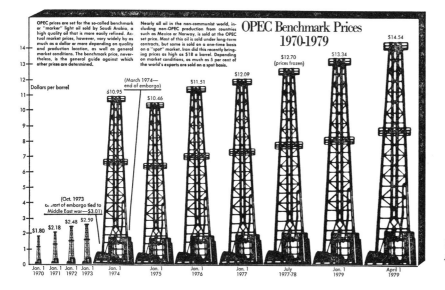

Washington Post, March 28, 1979, p. A-18.

All these accounts of oil prices made a second error, by showing the price of oil in inflated (current) dollars. The 1972 dollar was worth much more than the 1979 dollar. Thus in sweeping from left to right over the surface of the graphic, the vertical scale in effect changes—design variation—because the value of money changes over the years shown. The only way to think clearly about money over time is to make comparisons using inflation-adjusted units of money. Several distinguished graphic designers did express the price in real dollars—and they also avoided other sources of design variation. The stars were *Business Week*, the *Sunday Times* (London), and *The Economist*.

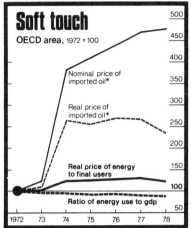

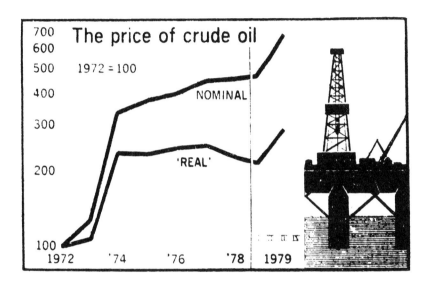

The Economist, December 29, 1979, p. 41.

Sunday Times (London), December 16, 1979, p. 54.

Business Week, April 9, 1979, p. 99.

In the graphic we saw first, the two sources of design variation covered up an intriguing, non-obvious aspect of the data: in the four years prior to the 1979–1980 increases, the real price of oil had *declined*. Busy with decoration, the graphic had missed the news.

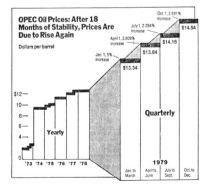

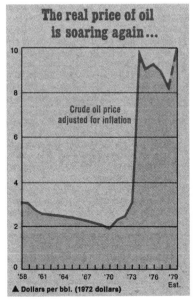

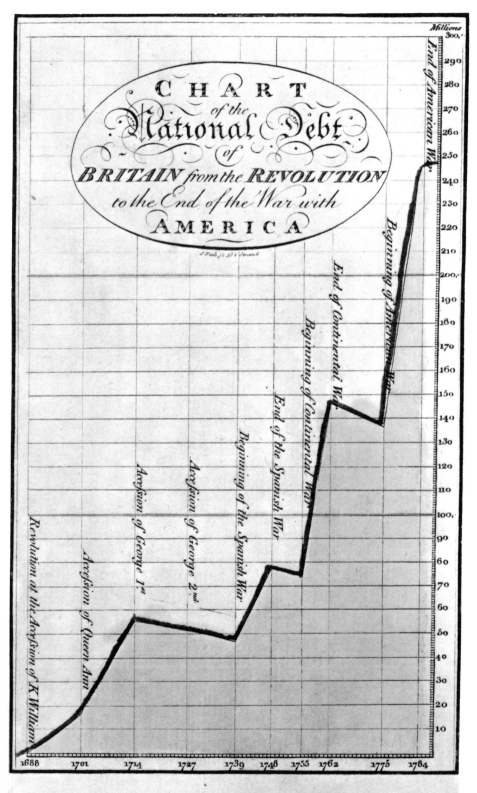

The Divisions at the Bottom are Years, & those on the Right hand Money.

The Case of Skyrocketing Government Spending

Probably the most frequently printed graphic, other than the daily weather map and stock-market trend line, is the display of government spending and debt over the years. These arrays nearly always create the impression that spending and debt are rapidly increasing.

As usual, Playfair was the first, publishing this finely designed graphic in 1786. Accompanied by his polemic against the "ruinous folly" of the British government policy of financing its colonial wars through debt, it is surely the first skyrocketing government debt chart, beginning the now 200-year history of such displays. This is one of the few Playfairs that is taller than wide; less than one-tenth of all his graphics (about 90, drawn during 35 years of work) are longer on the vertical. The tall shape here serves to emphasize the picture of rapid growth. The money figures are not adjusted for inflation.

But Playfair had the integrity to show an alternative version a few pages later in *The Commercial and Political Atlas*. The interest on the national debt was plotted on a broad horizontal scale, diminishing the skyrocket effect. And, furthermore, "This is in real and not in nominal millions" (page 129):

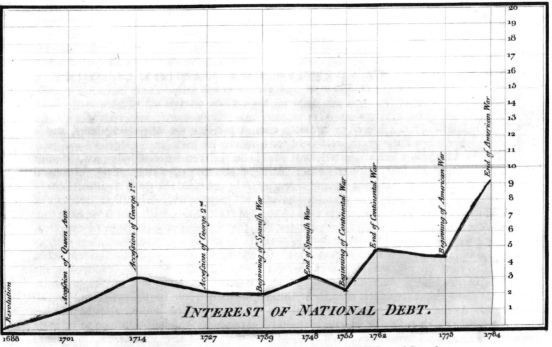

Interest of the NATIONAL DEBT from the Revolution.

The Bottom line is Years, those on the Right hand Millions of Pounds.

Although Playfair deflated money units over time in his work of
1786, the matter has proved difficult for many, eluding even modern scholars. This display helps its political point along by failing
to discount for inflation and population growth and by using a
tall and thin shape (the area covered by the data is 2.7 times taller
than wide):

Morris Fiorina, *Congress: Keystone of the
Washington Establishment* (New Haven,
1977), p. 92.

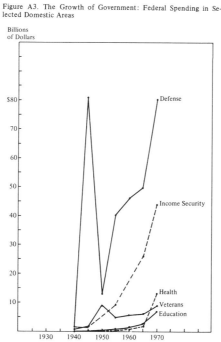

Figure A3. The Growth of Government: Federal Spending in Selected Domestic Areas

Let us look, in detail, at another graphic on government spending:

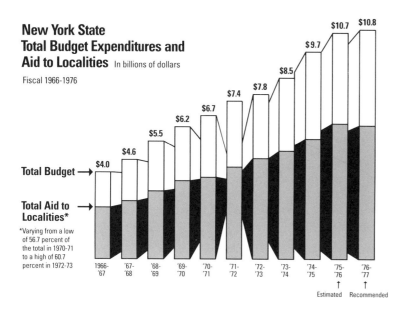

New York State
Total Budget Expenditures and
Aid to Localities In billions of dollars

Fiscal 1966-1976

*Varying from a low
of 56.7 percent of
the total in 1970-71
to a high of 60.7
percent in 1972-73

New York Times, February 1, 1976, p.
IV-6.

Despite the appearance created by the hyperactive design, the state budget actually did not increase during the last nine years shown. To generate the thoroughly false impression of a substantial and continuous increase in spending, the chart deploys several visual and statistical tricks—all working in the same direction, to exaggerate the growth in the budget. These graphical gimmicks:

These three parallelepipeds have been placed on an optical plane *in front* of the other eight, creating the image that the newer budgets tower over the older ones.

This cluster of type emphasizes and stretches out the low value for 1966–1967, encouraging the impression that recent years have shot up from a small, stable base. Horizontal arrows provide similar emphasis.

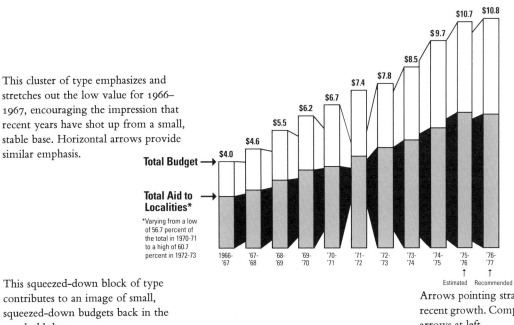

This squeezed-down block of type contributes to an image of small, squeezed-down budgets back in the good old days.

Arrows pointing straight up emphasize recent growth. Compare with horizontal arrows at left.

Leaving behind the distortion in the chartjunk heap at the left yields a calmer view:

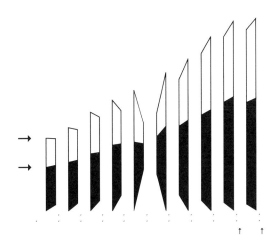

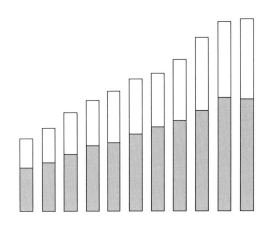

Two statistical lapses also bias the chart. First, during the years shown, the state's population increased by 1.7 million people, or 10 percent. Part of the budget growth simply paralleled population growth. Second, the period was a time of substantial inflation; those goods and services that cost state and local governments $1.00 to purchase in 1967 cost $2.03 in 1977. By not deflating, the graphic mixes up changes in the value of money with changes in the budget.

Application of arithmetic makes it possible to take population and inflation into account. Computing expenditures in *constant (real) dollars per capita* reveals a quite different—and far more accurate—picture:

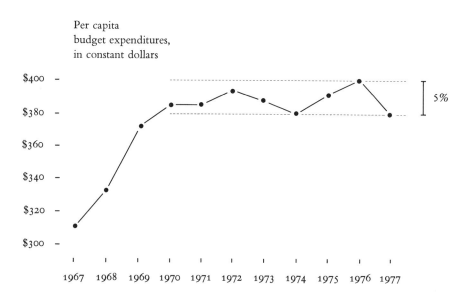

Thus, in terms of real spending per capita, the state budget increased by about 20 percent from 1967 to 1970 and remained relatively constant from 1970 through 1976. And the 1977 budget represents a substantial *decline* in expenditures. That is the real news story of these data, and it was completely missed by the Graph of the Magical Parallelepipeds. Of course no small set of numbers is going to capture the complexities of a large budget—but, at any rate, why tell lies?

The principle:

> In time-series displays of money, deflated and standardized units of monetary measurement are nearly always better than nominal units.

Visual Area and Numerical Measure

Another way to confuse data variation with design variation is to use areas to show one-dimensional data:

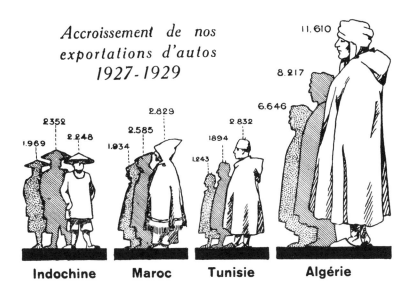

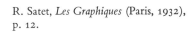
R. Satet, *Les Graphiques* (Paris, 1932), p. 12.

And here is the incredible shrinking doctor, with a Lie Factor of 2.8, not counting the additional exaggeration from the overlaid perspective and the incorrect horizontal spacing of the data:

Los Angeles Times, August 5, 1979, p. 3.

Many published efforts using areas to show magnitudes make the elementary mistake of varying both dimensions simultaneously in response to changes in one-dimensional data. Typical is the shrinking dollar fallacy. To depict the rate of inflation, graphs show currency shrinking on two dimensions, even though the value of money is one-dimensional. Here is one of hundreds of such charts:

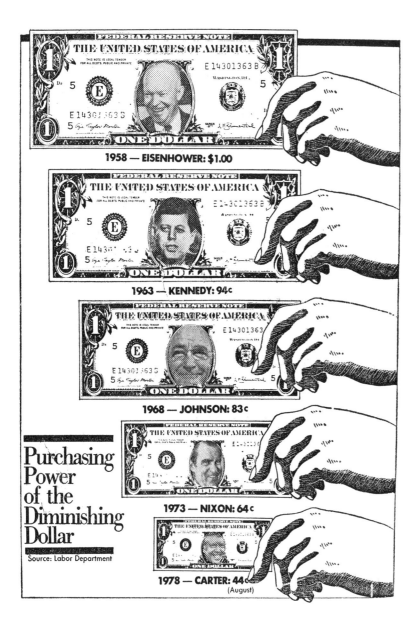

Washington Post, October 25, 1978, p. 1.

If the area of the dollar is accurately to reflect its purchasing power, then the 1978 dollar should be about twice as big as that shown.

There are considerable ambiguities in how people perceive a two-dimensional surface and then convert that perception into a one-dimensional number. Changes in physical area on the surface of a graphic do not reliably produce appropriately proportional changes in perceived areas. The problem is all the worse when the areas are tricked up into three dimensions:

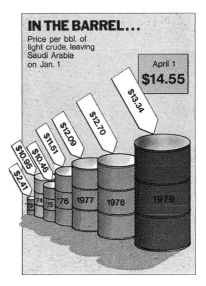

By surface area, the Lie Factor for this graphic is 9.4. But, if one takes the barrel metaphor seriously and assumes that the *volume* of the barrels represents the price change, then the volume from 1973 to 1979 increases 27,000 percent compared to a data increase of 454 percent, for a Lie Factor of 59.4, which is a record.

Similarly, a three-dimensional representation puffing up one-dimensional data:

New York Times, January 27, 1981, p. D–1.

Conclusion: The use of two (or three) varying dimensions to show one-dimensional data is a weak and inefficient technique, capable of handling only very small data sets, often with error in design and ambiguity in perception. These designs cause so many problems that they should be avoided:

> The number of information-carrying (variable) dimensions depicted should not exceed the number of dimensions in the data.

CASSE POSTALI DI RISPARMIO ITALIANE

Numero dei Libretti, Libretto medio e Deposito totale

al fine di ogni mese

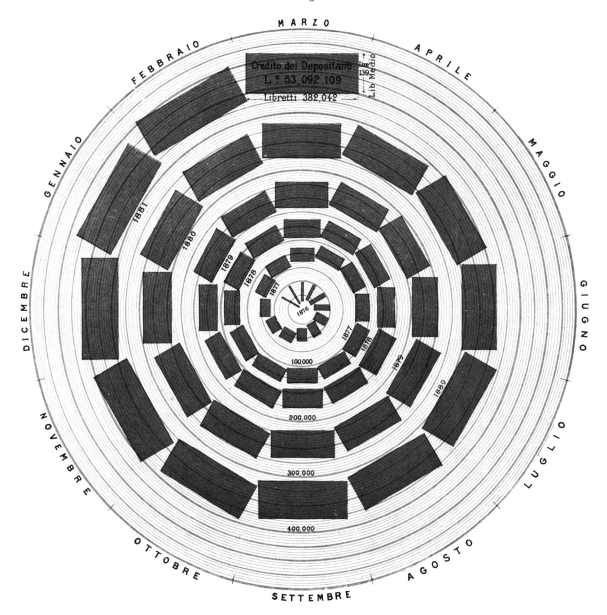

This multivariate history of the Italian post office uses two dimensions in a way nearly consistent with this principle, with the number of postal savings books issued and the average size of deposits multiplying up to total deposits at the end of each month from 1876 to 1881.

Antonio Gabaglio, *Teoria Generale della Statistica* (Milan, second edition, 1888).

But Playfair's circles, an early use of area to show magnitude, are not consistent with the principle, since the one-dimensional data (city populations) are represented by an areal data measure:

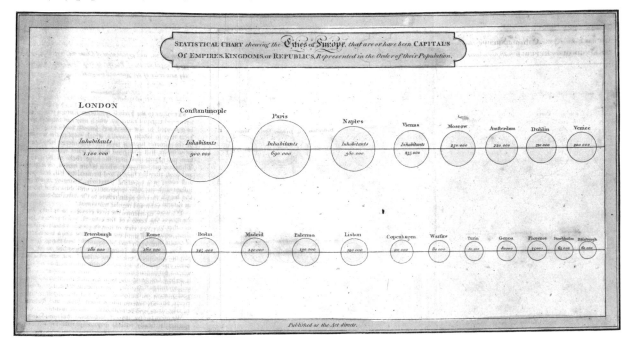

Perhaps graphics that border on cartoons should be exempt from the principle. We certainly would not want to forgo the 4,340 pound chicken:

Scientific American Reference Book (New York, 1909), p. 280.

Context is Essential for Graphical Integrity

To be truthful and revealing, data graphics must bear on the question at the heart of quantitative thinking: "Compared to what?" The emaciated, data-thin design should always provoke suspicion, for graphics often lie by omission, leaving out data sufficient for comparisons. The principle:

Graphics must not quote data out of context.

Nearly all the important questions are left unanswered by this display:

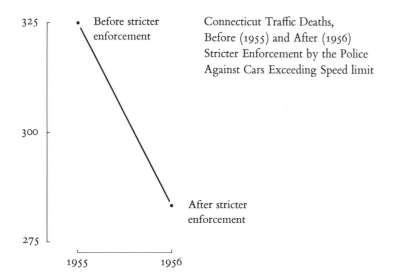

A few more data points add immensely to the account:

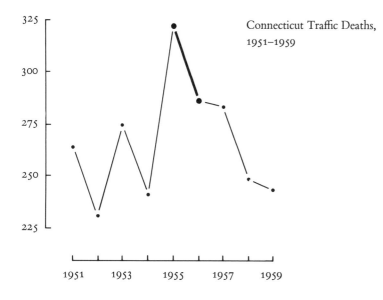

Imagine the very different interpretations other possible time-paths surrounding the 1955–1956 change would have:

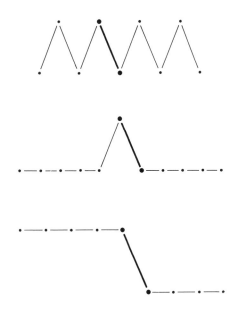

Comparisons with adjacent states give a still better context, revealing it was not only Connecticut that enjoyed a decline in traffic fatalities in the year of the crackdown on speeding:

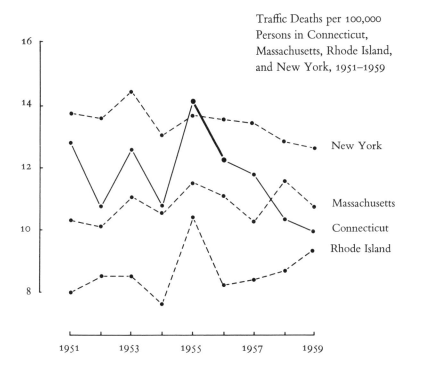

Traffic Deaths per 100,000 Persons in Connecticut, Massachusetts, Rhode Island, and New York, 1951–1959

Donald T. Campbell and H. Laurence Ross, "The Connecticut Crackdown on Speeding: Time Series Data in Quasi-Experimental Analysis," in Edward R. Tufte, ed., *The Quantitative Analysis of Social Problems* (Reading, Mass., 1970), 110–125.

Conclusion

Lying graphics cheapen the graphical art everywhere. Since the lies often show up in news reports, millions of images are printed. When a chart on television lies, it lies tens of millions of times over; when a *New York Times* chart lies, it lies 900,000 times over to a great many important and influential readers. The lies are told about the major issues of public policy—the government budget, medical care, prices, and fuel economy standards, for example. The lies are systematic and quite predictable, nearly always exaggerating the rate of recent change.

   The main defense of the lying graphic is . . . "Well, at least it was approximately correct, we were just trying to show the general direction of change." But many of the deceptive displays we saw in this chapter involved fifteenfold lies, too large to be described as approximately correct. And in several cases the graphics were not even approximately correct by the most lax of standards, since they falsified the real news in the data. It is the special character of numbers that they have a magnitude as well as an order; numbers measure *quantity*. Graphics can display the quantitative size of changes as well as their direction. The standard of getting only the direction and not the magnitude right is the philosophy that informs the Pravda School of Ordinal Graphics. There, every chart has a crystal clear direction coupled with fantasy magnitudes.

Рост продукции промышленности (1922 г. = 1).

Pravda, May 24, 1982, p. 2.

A second defense of the lying graphic is that, although the design itself lies, the actual numbers are printed on the graphic for those picky folks who want to know the correct size of the effects displayed. It is as if not lying in one place justified fifteenfold lies elsewhere. Few writers would work under such a modest standard of integrity, and graphic designers should not either.

Graphical integrity is more likely to result if these six principles are followed:

> The representation of numbers, as physically measured on the surface of the graphic itself, should be directly proportional to the numerical quantities represented.

> Clear, detailed, and thorough labeling should be used to defeat graphical distortion and ambiguity. Write out explanations of the data on the graphic itself. Label important events in the data.

> Show data variation, not design variation.

> In time-series displays of money, deflated and standardized units of monetary measurement are nearly always better than nominal units.

> The number of information-carrying (variable) dimensions depicted should not exceed the number of dimensions in the data.

> Graphics must not quote data out of context.

3 *Sources of Graphical Integrity and Sophistication*

Why do artists draw graphics that lie? Why do the world's major newspapers and magazines publish them?[1]

Although bias and stereotyping are the origin of more than a few graphical distortions, the primary causes of inept graphical work are to be found in the skills, attitudes, and organizational structure prevailing among those who design and edit statistical graphics.

Lack of Quantitative Skills of Professional Artists

Lurking behind the inept graphic is a lack of judgment about quantitative evidence. Nearly all those who produce graphics for mass publication are trained exclusively in the fine arts and have had little experience with the analysis of data. Such experience is essential for achieving precision and grace in the presence of statistics, but even textbooks of graphical design are silent on how to think about numbers. Illustrators too often see their work as an exclusively artistic enterprise—the words "creative," "concept," and "style" combine regularly in all possible permutations, a Big Think jargon for the small task of constructing a time-series a few data points long. Those who get ahead are those who beautify data, never mind statistical integrity.

The Doctrine That Statistical Data Are Boring

Inept graphics also flourish because many graphic artists believe that statistics are boring and tedious. It then follows that decorated graphics must pep up, animate, and all too often exaggerate what evidence there is in the data. For example:

- *Time*'s first full-time chart specialist, an art-school graduate, says that in his work, "The challenge is to present statistics as a visual idea rather than a tedious parade of numbers."[2]

- The opening sentence of the chapter on statistical charts in Jan White's *Graphic Idea Notebook*: "Why are statistics so boring?" Sample illustrations supposedly reveal "Dry statistics turned

[1] "It is difficult to know why these same errors are being repeated. In Playfair's original work these kinds of mistakes were not made; moreover, these errors were not as widespread in the 1930's as they are now. Perhaps the reason is an increase in the perceived need for graphs . . . without a concomitant increase in training in their construction. Evidence gathered by the committee on graphics of the American Statistical Association indicates that formal training in graphic presentation has had a marked decline at all levels of education over the last few decades." Howard Wainer, "Making Newspaper Graphs Fit to Print," in Paul A. Kolers, et al., eds. *Processing of Visible Language 2* (New York, 1980), p. 139.

[2] *Time*, February 11, 1980, p. 3.

into symbolic graphics" and "Plain statistics embellished or humanized with pictures."[3]

- A fine book on graphics, Herdeg's *Graphis/Diagrams*, is described by its publisher: "An international review demonstrating convincingly that statistical and diagrammatic graphics do not necessarily have to be dull."[4]

The doctrine of boring data serves political ends, helping to advance certain interests over others in bureaucratic struggles for control of a publication's resources. For if the numbers are dull dull dull, then an artist, indeed many artists, indeed an Art Department and an Art Director are required to animate the data, lest the eyes of the audience glaze over. Thus the doctrine encourages placing data graphics under control of artists rather than in the hands of those who write the words and know the substance. As the art bureaucracy grows, style replaces content. And the word people, having lost space in the publication to data decorators, console themselves with thoughts that statistics are really rather tedious anyway.

If the statistics are boring, then you've got the wrong numbers. Finding the right numbers requires as much specialized skill—statistical skill—and hard work as creating a beautiful design or covering a complex news story.

The Doctrine That Graphics Are Only for the Unsophisticated Reader

Many believe that graphical displays should divert and entertain those in the audience who find the words in the text too difficult. For example:

- *Consumer Reports* describes the design of their new consumer magazine for children: "For the first test issue, CU's professional staff produced an article about sugar that was longer on graphics than on information. We had feared children might be overwhelmed by too many facts."[5]

- An art director with overall responsibility for the design of some 3,000 data graphics each year (yielding 2.5 billion printed images) said that graphics are intended more to lure the reader's attention away from the advertising than to explain the news in any detail. "Unlike the advertisements," he said, "at least we don't put naked women in our graphics."[6]

[3] Jan V. White, *Graphic Idea Notebook* (New York, 1980), pp. 148, 165.

[4] Walter Herdeg, ed., *Graphis/Diagrams* (Zurich, 1976).

[5] *Consumer Reports*, 45 (July 1980), 408.

[6] Louis Silverstein, "Graphics at the *New York Times*," presentation at the First General Conference on Social Graphics, Leesburg, Virginia, October 23, 1978.

- A news director at a national television network said that graphics must be instantly understandable: "If you have to explain it, don't use it."[7]

[7] Interview with author, July 1980.

This kind of graphical thinking leads to

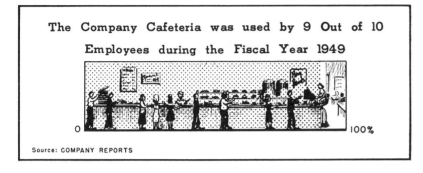

Mary Eleanor Spear, *Charting Statistics* (New York, 1952), p. 5, who appropriately describes this as an "unnecessary chart."

The Consequences

What E. B. White said of writing is also true of statistical graphics: "No one can write decently who is distrustful of the reader's intelligence, or whose attitude is patronizing."[8] Contempt for graphics and their audience, along with the lack of quantitative skills among illustrators, has deadly consequences for graphical work: over-decorated and simplistic designs, tiny data sets, and big lies.

[8] In William Strunk, Jr., and E. B. White, *The Elements of Style* (New York, 1959), p. 70.

Like censorship, these constraints on graphical design lead to elliptical and eccentric communication. In seeking to avoid the subtleties of the scatterplot, artists drew up these convoluted specimens, forcing bivariate data into a univariate design:

New York Times, June 16, 1980, p. A-18.

Allen D. Manvel, "Taxation and Economic Growth," *Taxation with Representation Newsletter*, 9 (June 1980), p. 3.

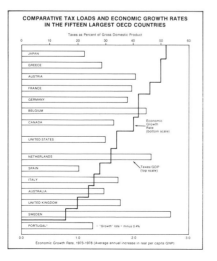

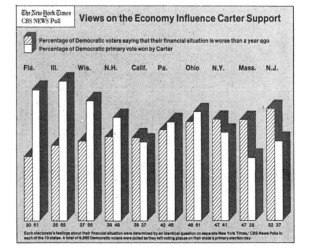

But beyond reviewing a few examples, let us look more systematically at the level of graphical sophistication prevailing at different publications. In order to make comparisons among a variety of newspapers, magazines, scientific journals, and books, I have compiled a rough measure of graphical sophistication—the share of a publication's graphics that are *relational*. Such a design links two or more variables but is not a time-series or a map. Relational graphics are essential to competent statistical analysis since they confront statements about cause and effect with evidence, showing how one variable affects another. The design idea is a simple one, although not quite as simple as the bar chart, time-series plot, or data map. Relational graphics have been used since 1765 and are printed billions of times and ways every year; and there is evidence that twelve-year-old children understand the design.[9]

All these graphics count as sophisticated by our hardly demanding measure:[10]

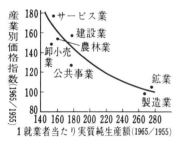

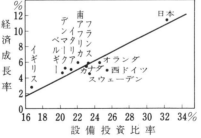

The frequency of use of relational designs was counted for randomly selected issues from 1974 to 1980 of each of 15 news publications. A total of about 4,000 graphics were examined in sampled issues. Scaling up the observed data by the frequency and circulation of the publication indicates that the sample represents a population of 250 to 300 billion printed graphical images.

[9] Clara Francis Bamberger, "Interpretation of Graphs at the Elementary School Level," *Catholic University of America Educational Research Monographs*, 13 (May 1942). Additional data from textbooks and standardized tests are presented shortly.

[10] A variety of measures of graphical intelligence and complexity are possible and another, data density, is discussed in Chapter 8.

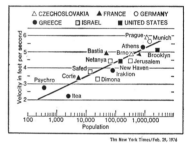

The New York Times/Feb. 29, 1976

Pace of City Life Found 2.8 Feet per Second Faster

By BOYCE RENSBERGER

The pace of life in big cities is faster than it is in small towns—about 2.8 feet per second faster, according to a study by a Princeton University psychologist and his wife, who is an anthropologist.

By measuring how fast people walk along the main streets of municipalities of varying sizes, they have confirmed what most people have sensed informally. The bigger the city, the faster its inhabitants walk.

They found, for example, that on Flatbush Avenue in Brooklyn, people walk at a brisk 5 feet per second, only a little slower than their counterparts on Wenceslas Square in Prague, who bustle along at 5.8 feet per second.

In contrast to Brooklyn and Prague, both of which have a population of more than a million, the 365 citizens of Psychro, Greece, amble along at 2.7 feet per second and the people of Corte, France (population 5,500, move at 3.3 feet per second.

New York Times, February 29, 1976, p. 46.

Isao Sato and Miyohei Shinohara, *New Politics and Economics* (Tokyo, 1974), p. 113; a Japanese high school textbook.

Table 1 shows the results, ranking the 15 news publications by graphical sophistication. Seven of the papers, from *Pravda* to the *Wall Street Journal*, produced no relational graphics among those sampled and usually limited themselves to time-series. Other papers published more advanced graphics: the Japanese *Asahi* (a mass circulation daily), *Akahata* ("Red Flag," a Communist party paper that appears, from the data, to have employed a sophisticated and talented graphical designer in 1979), and *Nihon Keizai* (a financial daily), as well as *Der Spiegel* and *The Economist*. Although none reached the level of sophistication found in displays of scientific data (a random sample of 220 graphics from *Science* 1978–1980 had 42 percent of relational design), it is clear that some graphical intelligence is possible in news work, at least in Japan and at a few European weeklies.

Table 1
Graphical Sophistication, World Press, 1974–1980

	Percentage of statistical graphics based on more than one variable, but not a time-series or a map	Number of graphics in sample
Akahata ("Red Flag") (Japan, daily, circulation 30,000)	9.3%	202
Asahi Shimbun (Japan, daily, 8,000,000)	7.6%	119
Der Spiegel (Germany, weekly, 1,000,000)	5.7%	454
The Economist (Britain, weekly, 170,000)	2.0%	342
Nihon Keizai Shimbun (Japan, daily financial paper, 1,700,000)	1.7%	297
Le Monde (French, daily, 440,000)	0.7%	144
Business Week (U.S., weekly, 800,000)	0.6%	726
New York Times (U.S., daily, 900,000; Sunday, 1,500,000)	0.5%	422
Pravda (USSR, daily, 10,500,000)	0.0%	54
Frankfurter Allgemeine (Germany, daily, 300,000)	0.0%	93
The Times (Britain, daily, 400,000)	0.0%	107
Washington Post (U.S., daily, 600,000; Sunday, 800,000)	0.0%	121
Time (U.S., weekly, 4,300,000)	0.0%	147
Die Zeit (Germany, weekly, 300,000)	0.0%	213
Wall Street Journal (U.S., daily, 2,000,000)	0.0%	449

Japanese graphical distinction is consistent with that country's heavy use of statistical techniques in the workplace and extensive quantitative training, even in the early years of school:

> . . . no nation ranks higher in its collective passion for statistics. In Japan, statistics are the subject of a holiday, local and national conventions, awards ceremonies and nationwide statistical collection and graph-drawing contests. "This year," said Yoshiharu Takahashi, a Government statistician, "we had almost 30,000 entries. Actually, we had 29,836."
>
> Entries in the [children's] statistical graph contest were screened three times by judges, who gave first prize this year to the work of five 7-year-olds. Their graph creation, titled "Mom, play with us more often," was the result of a survey of 32 classmates on the frequency that mothers play with their offspring and the reasons given for not doing so. . . . Other children's work examined the frequency of family phone usage and correlated the day's temperature with cicada singing.[11]

Note the relational design of the last children's graphic mentioned.

The five U.S. publications examined rank toward the bottom of the world list, along with *Pravda* and a few European papers. Note, in Table 1, the complete dominance of non-relational designs at the lower-ranked newspapers and magazines. This is unfortunate because the relational graphic, unlike the simpler designs, is an *explanatory* graphic—surely a natural for news reporting and analysis.

The statistical graphics found in college and even high school textbooks are more sophisticated than those in news publications. Indeed, grade school children may experience a greater density of relational graphics than someone who reads only *Business Week*, the *New York Times*, *Time*, the *Wall Street Journal*, and the *Washington Post*. Tables 2 and 3 record the graphical sophistication of textbooks and of a variety of standardized educational tests. A comparison between these data and Table 1 suggests that most news publications outside of Japan operate at a pre-adult level of intelligence in graphical design.[12]

[11] Andrew H. Malcolm, "Data-Loving Japanese Rejoice on Statistics Day," *New York Times*, October 28, 1977, p. A-1. See especially James R. Beniger and D. Eleanor Westney, "Japanese and U.S. Media: Graphics as a Reflection of Newspapers' Social Role," *Journal of Communication*, 31 (Spring 1981), 14–27, which compares *Asahi* and the *New York Times*. The failure of the Japanese to be bored by statistics accounts in part for their superior economic and industrial performance; see Robert E. Cole, *Work, Mobility, and Participation* (Berkeley, 1980).

[12] Readers of news publications, particularly the elite press, have considerable educational and professional attainments, with the resulting graphical skills. About 80 percent of the 1.5 million readers of the Sunday *New York Times* attended college, according to a 1980 *Times* market survey. The audience for statistical graphics is smarter than many illustrators believe.

Table 2
Graphical Sophistication, College and High School Textbooks

	Percentage of statistical graphics based on more than one variable, but not a time-series or a map	Number of graphics
COLLEGE TEXTBOOKS:		
Medicine and public health: 11 articles in Judith Tanur, et al., *Statistics: A Guide to the Unknown*	82%	17
Introduction to Psychology, by Ernest Hilgard, et al.	68%	82
General Chemistry, by Linus Pauling	66%	53
Life on Earth, by Edward Wilson, et al.	47%	59
Weather, astronomy, engineering: 7 articles in Tanur, *Statistics: A Guide to the Unknown*	44%	9
Communication, work, education, economics: 20 articles in Tanur, *Statistics: A Guide to the Unknown*	43%	35
Political Behavior of the American Electorate, by William H. Flanigan and Nancy H. Zingale	42%	43
Economics, by Paul Samuelson	16%	57
Democracy in America, by Robert A. Dahl	8%	25
American Government, by James Q. Wilson	0%	39
HIGH SCHOOL TEXTBOOKS:		
Chemical Principles, by William Masterton and Emil Slowinski	77%	27
The Project Physics Course, by Harvard Project Physics	48%	33
New Politics and Economics, by Isao Sato and Miyohei Shinohara (Japanese)	27%	22
Biological Science: An Ecological Approach, Biological Sciences Curriculum Study	18%	28
The American Economy, by Roy J. Sampson, et al.	5%	132
Sociology: The Study of Human Relationships, by LaVerne Thomas and Robert Anderson	0%	3
New Ethics and Social Science, by Yokichi Yajima, et al. (Japanese)	0%	5
Rise of the American Nation, by Lewis Paul Todd and Merle Curti	0%	39
Magruder's American Government, revised by William McClenaghan	0%	70

Table 3
Graphical Sophistication, Educational Tests

	Percentage of statistical graphics based on more than one variable, but not a time-series or a map	Number of graphics
National university entrance examinations, Japan, 1979 and 1980	100%	16
Review materials, Law School Admission Test, United States 1975	48%	29
Standardized tests for grade school, high school, and college; United States, 1970s:*		
Science, 14 tests	67%	64
Arithmetic, mathematics, algebra, and analytic geometry; 21 tests	41%	37
Social studies, history, and government; 14 tests	24%	49
General ability, 5 tests	21%	47

*Graphics collected in James R. Beniger, compiler, *Selected Standardized Test Items that Measure Ability with Graphics* (Washington, D.C.: Bureau of Social Science Research, 1975).

And so, just as there is a double standard of integrity at a good many news publications—one for words, another for graphics— so there is a double standard of sophistication. The statistical graphics are stupid; the prose is often serious and sometimes even demanding of expertise, as can be seen in these sentences from a single issue of the *New York Times*:

> Recycling petrodollars may postpone the day of reckoning, but its effects would soon become intolerable without a steady depreciation in their purchasing power. Floating rates of exchange cannot restore even a semblance of equilibrium.

> Numerous facets of the performance seem decidedly unfashionable if not downright eccentric: the square-toed instrumental phrasing and the frequent plodding tempos in the arias, the Romanticized treatment of the chorales, the generous retards at every cadence, the often intrusively elaborate continuo improvisations and an inconsistent attitude toward expression which ranges from heaving Mahlerian emphases to mechanical literalism.

> The Court shows no sign of retreating from its view that a state government is protected by sovereign immunity against court orders to pay retroactive damages for past violations.

> And Dr. Garth Graham, a medical director with Smithkline Corp., makers of Thorazine, noted that neuroleptics produce no euphoria, and are therefore unlikely to be abused by patients with a history of drug or alcohol dependence. "They are, if anything, dysphorogenic," Dr. Graham said.

Conclusion

The conditions under which many data graphics are produced—
the lack of substantive and quantitative skills of the illustrators,
dislike of quantitative evidence, and contempt for the intelligence
of the audience—guarantee graphic mediocrity. These conditions
engender graphics that (1) lie; (2) employ only the simplest designs,
often unstandardized time-series based on a small handful of
data points; and (3) miss the real news actually in the data.

It wastes the tremendous communicative power of graphics to
use them merely to decorate a few numbers. Moreover, much of
the world these days is observed and assessed quantitatively—and
well-designed graphics are far more effective than words in
showing such observations.

How can graphic mediocrity be remedied?

Surely there is something to be said for rejecting once and for
all the doctrines that data graphics are for the unintelligent and
that statistics are boring. These doctrines blame the victims (the
audience and the data) rather than the perpetrators.

Graphical competence demands three quite different skills: the
substantive, statistical, and artistic. Yet now most graphical work,
particularly at news publications, is under the direction of but a
single expertise—the artistic. Allowing artist-illustrators to control
the design and content of statistical graphics is almost like allowing
typographers to control the content, style, and editing of prose.
Substantive and quantitative expertise must also participate in the
design of data graphics, at least if statistical integrity and graphical
sophistication are to be achieved.

PART II
Theory of Data Graphics

Everyone spoke of an information overload, but what there was in fact was a non-information overload.

Richard Saul Wurman, *What-If, Could-Be* (Philadelphia, 1976)

4 Data-Ink and Graphical Redesign

Data graphics should draw the viewer's attention to the sense and substance of the data, not to something else. The data graphical form should present the quantitative contents. Occasionally artfulness of design makes a graphic worthy of the Museum of Modern Art, but essentially statistical graphics are instruments to help people reason about quantitative information.

Playfair's very first charts devoted too much of their ink to graphical apparatus, with elaborate grid lines and detailed labels. This time-series, engraved in August 1785, is from the early pages of *The Commercial and Political Atlas*:

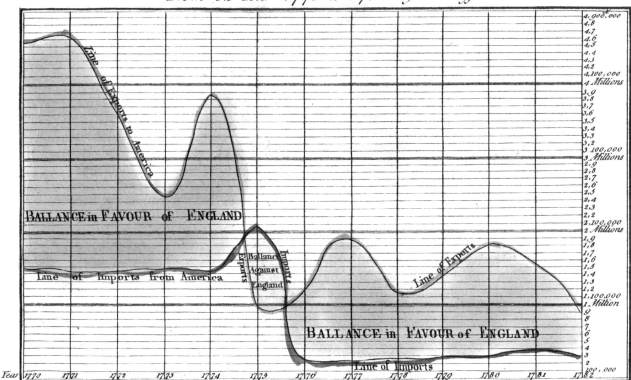

The Bottom Line is divided into Years the right-hand Line into HUNDRED THOUSAND POUNDS

Within a year Playfair had eliminated much of the non-data detail in favor of cleaner design that focused attention on the time-series itself. He then began working with a new engraver and was soon producing clear and elegant displays:

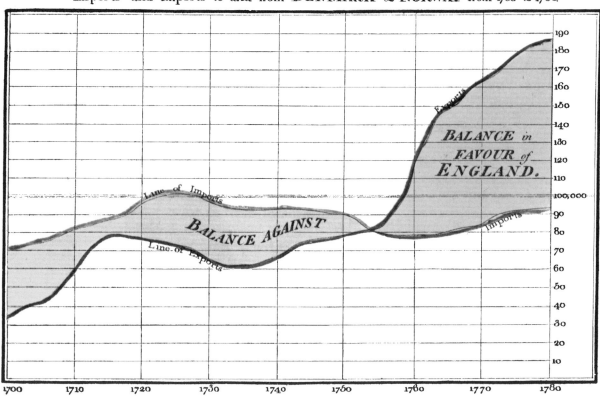

Exports and Imports to and from DENMARK & NORWAY from 1700 to 1780.

The Bottom line is divided into Years, the Right hand line into L10,000 each.

Published as the Act directs, 1st May 1786, by Wm Playfair   Neele sculpt 352, Strand, London.

This improvement in graphical design illustrates the fundamental principle of good statistical graphics:

> Above all else show the data.

The principle is the basis for a theory of data graphics.

Data-Ink

A large share of ink on a graphic should present data-information, the ink changing as the data change. *Data-ink* is the non-erasable core of a graphic, the non-redundant ink arranged in response to variation in the numbers represented. Then,

$$\text{Data-ink ratio} = \frac{\text{data-ink}}{\text{total ink used to print the graphic}}$$

= proportion of a graphic's ink devoted to the non-redundant display of data-information

= 1.0 − proportion of a graphic that can be erased without loss of data-information.

A few graphics use every drop of their ink to convey measured quantities. Nothing can be erased without losing information in these continuous eight tracks of an electroencephalogram. The data change from background activity to a series of polyspike bursts. Note the scale in the bottom block, lower right:

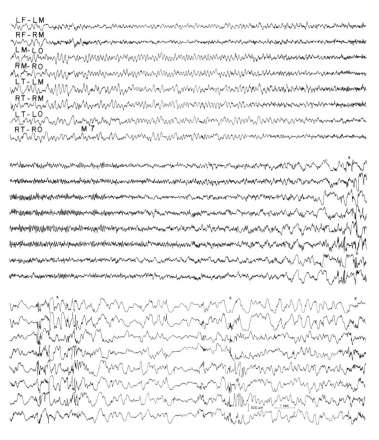

Kenneth A. Kooi, *Fundamentals of Electroencephalography* (New York, 1971), p. 110.

Most of the ink in this graphic is data-ink (the dots and labels on the diagonal), with perhaps 10–20 percent non-data-ink (the grid ticks and the frame):

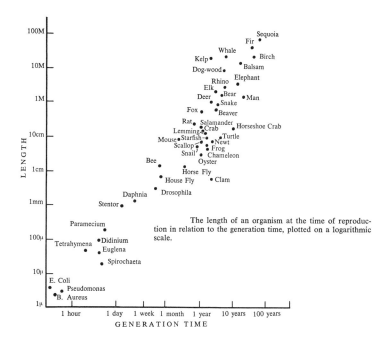

John Tyler Bonner, *Size and Cycle: An Essay on the Structure of Biology* (Princeton, 1965), p. 17.

In this display with nearly all its ink devoted to matters other than data, the grid sea overwhelms the numbers (the faint points scattered about the diagonal):

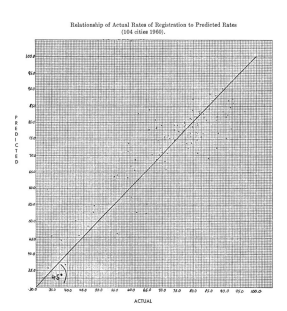

Another published version of the same data drove the share of data-ink up to about 0.7, an improvement:

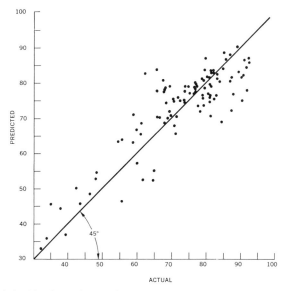

Relationship of Actual Rates of Registration to Predicted Rates (104 cities 1960).

But a third reprint publication of the same figure forgot to plot the points and simply retraced the grid lines from the original, including the excess strip of grid along the top and right margins. The resulting figure achieves a graphical absolute zero, a null data-ink ratio:

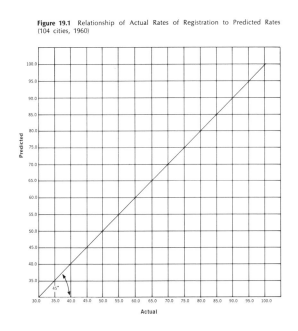

Figure 19.1 Relationship of Actual Rates of Registration to Predicted Rates (104 cities, 1960)

The three graphics were published in, respectively, Stanley Kelley, Jr., Richard E. Ayres, and William G. Bowen, "Registration and Voting: Putting First Things First," *American Political Science Review*, 61 (1967), 371; then reprinted in Edward R. Tufte, ed., *The Quantitative Analysis of Social Problems* (Reading, Mass., 1970), p. 267; and reprinted again in William J. Crotty, ed., *Public Opinion and Politics: A Reader* (New York, 1970), p. 364.

Maximizing the Share of Data-Ink

The larger the share of a graphic's ink devoted to data, the better (other relevant matters being equal):

Maximize the data-ink ratio, within reason.

Every bit of ink on a graphic requires a reason. And nearly always that reason should be that the ink presents new information.

The principle has a great many consequences for graphical editing and design. The principle makes good sense and generates reasonable graphical advice—for perhaps two-thirds of all statistical graphics. For the others, the ratio is ill-defined or is just not appropriate. Most important, however, is that other principles bearing on graphical design follow from the idea of maximizing the share of data-ink.

Two Erasing Principles

The other side of increasing the proportion of data-ink is an erasing principle:

Erase non-data-ink, within reason.

Ink that fails to depict statistical information does not have much interest to the viewer of a graphic; in fact, sometimes such non-data-ink clutters up the data, as in the case of a thick mesh of grid lines. While it is true that this boring ink sometimes helps set the stage for the data action, it is surprising, as we shall see in Chapter 7, how often the data themselves can serve as their own stage.

Redundant data-ink depicts the same number over and over. The labeled, shaded bar of the bar chart, for example,

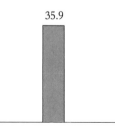

unambiguously locates the altitude in six separate ways (any five of the six can be erased and the sixth will still indicate the height): as the (1) height of the left line, (2) height of shading, (3) height of right line, (4) position of top horizontal line, (5) position (not content) of number at bar's top, and (6) the number itself. That is

more ways than are needed. Gratuitous decoration and reinforce-
ment of the data measures generate much redundant data-ink:

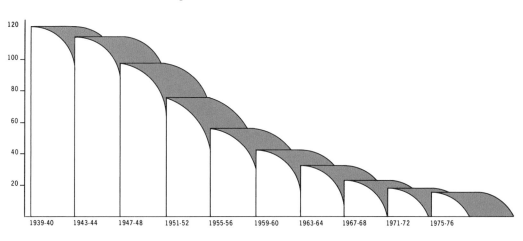

Bilateral symmetry of data measures also creates redundancy,
as in the box plot, the open bar, and Chernoff faces:

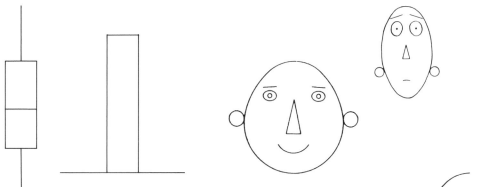

Half-faces carry the same information as full faces. Halves may
be easier to sort (by matching the right half of an unsorted face
to the left half of a sorted face) than full faces. Or else an
asymmetrical full face can be used to report additional variables.[1]

Bilateral symmetry doubles the space consumed by the design
in a graphic, without adding new information. The few studies
done on the perception of symmetrical designs indicate that "when
looking at a vase, for instance, a subject would examine one of its
symmetric halves, glance at the other half and, seeing that it was
identical, cease his explorations. . . . The enjoyment of symmetry
. . . lies not with the physical properties of the figure. At least eye
movements suggest anything but symmetry, balance, or rest."[2]

[1] Bernhard Flury and Hans Riedwyl,
"Graphical Representation of Multi-
variate Data by Means of Asymmetrical
Faces," *Journal of the American Statistical
Association*, 76 (December 1981), 757–
765.

[2] Leonard Zusne, *Visual Perception of
Form* (New York, 1970), pp. 256–257.

Redundancy, upon occasion, has its uses: giving a context and order to complexity, facilitating comparisons over various parts of the data, perhaps creating an aesthetic balance. In cyclical time-series, for example, parts of the cycle should be repeated so that the eye can track any part of the cycle without having to jump back to the beginning. Such redundancy possibly improves Marey's 1880 train schedule. Those people leaving Paris or Lyon in the evening find that their trains run off the right-hand edge of the chart, to be picked up on the left again:

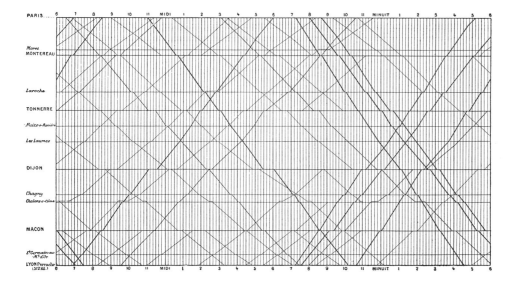

Attaching an extra half cycle makes every train in the first 24 hours of the schedule a continuous line (as would mounting the original on a cylinder):

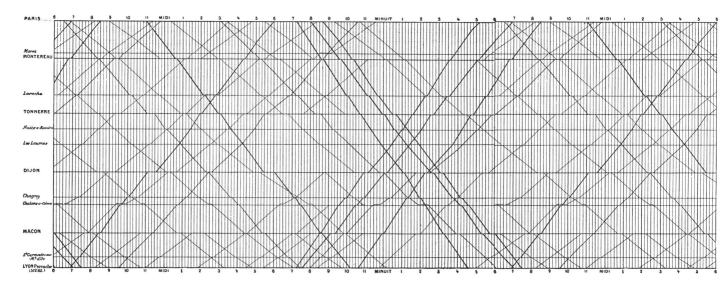

And, similarly, instead of once around the world in this display of surface ocean currents, one and two-thirds times around is better:

Kirk Bryan and Michael D. Cox, "The Circulation of the World Ocean: A Numerical Study. Part 1, A Homogeneous Model," *Journal of Physical Oceanography*, 2 (1972), 330.

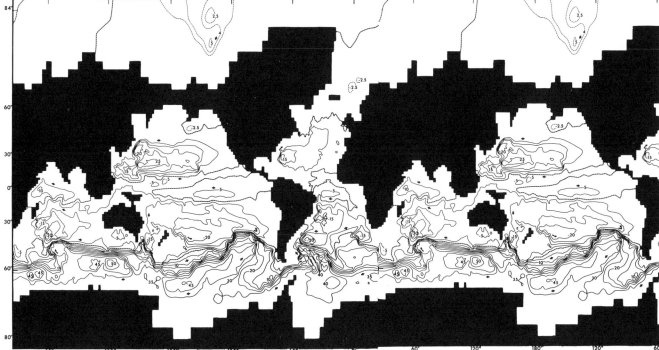

Most data representations, however, are of a single, uncomplicated number, and little graphical repetition is needed. Unless redundancy has a distinctly worthy purpose, the second erasing principle applies:

Erase redundant data-ink, within reason.

Application of the Principles in Editing and Redesign

Just as a good editor of prose ruthlessly prunes out unnecessary words, so a designer of statistical graphics should prune out ink that fails to present fresh data-information. Although nothing can replace a good graphical idea applied to an interesting set of numbers, editing and revision are as essential to sound graphical design work as they are to writing. T. S. Eliot emphasized the "capital importance of criticism in the work of creation itself. Probably, indeed, the larger part of the labour of an author in composing his work is critical labour; the labour of sifting, combining, constructing, expunging, correcting, testing: this frightful toil is as much critical as creative."[3]

Consider this display, which compares each long bar with the adjacent short bar to show the viewer that, under the various experimental conditions, the long bar is longer:

[3] T. S. Eliot, "The Function of Criticism," in *Selected Essays 1917–1932* (New York, 1932), p. 18.

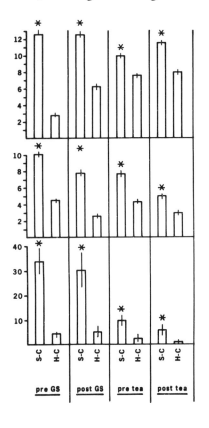

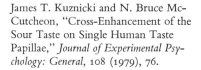

James T. Kuznicki and N. Bruce Mc-Cutcheon, "Cross-Enhancement of the Sour Taste on Single Human Taste Papillae," *Journal of Experimental Psychology: General*, 108 (1979), 76.

Vigorous pruning improves the graphic immensely, while still
retaining all the data of the original. It is remarkable that erasing
alone can work such a transformation:

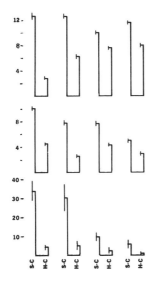

The horizontals indicate the paired comparisons and would change
if the experimental design changed—so they count as information-
carrying. All the asterisks are out since every paired comparison
was statistically significant, a point that the caption can note. Here
is the mix of non-data-ink and redundant data-ink that was erased,
about 65 percent of the original:

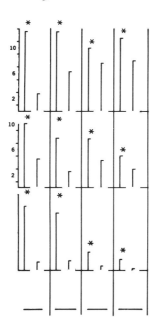

The data graphical arithmetic looks like this—the original design equals the erased part plus the good part:

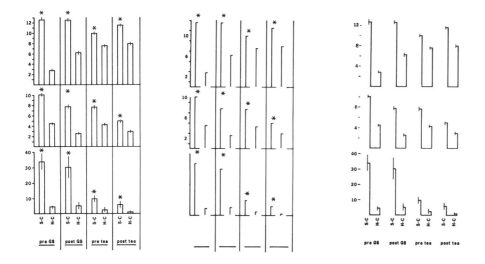

The next graphic, drawn by the distinguished science illustrator Roger Hayward, shows the periodicity of properties of chemical elements, exemplified by atomic volume as a function of atomic number. The data-ink ratio is less than 0.6, lowered because the 76 data points and the reference curves are obscured by the 63 dark grid marks arrayed over the data plane like a precision marching band of 63 mosquitoes:

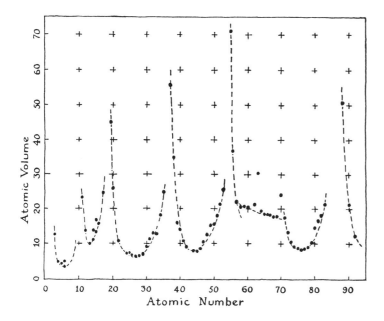

Linus Pauling, *General Chemistry* (San Francisco, 1947), p. 64.

The grid ticks compete with the essential information of the graphic, the curves tracing out the periods and the empirical observations. The little grid marks and part of the frame can be safely erased, removed from the denominator of the data-ink ratio:

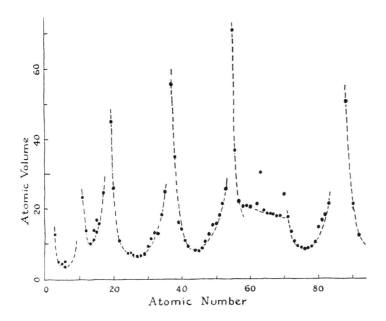

The uncluttered display brings out another aspect of the data: several of the elements do not fit the smooth theoretical curves all that well. The data-ink ratio has increased to about .9, with only the frame lines remaining as pure non-information:

The reference curves prove essential for organizing the data to
show the periodicity. The curves create a structure, giving an
ordering, a hierarchy, to the flow of information from the page:

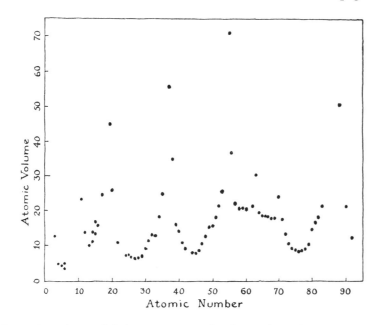

Restoring the grid fails to organize the data. The ticks are too
powerful, and they also add a disconcerting visual vibration to the
graphic. With the ticks, the reference curves become all the more
necessary, since the eye needs some guidance through the maze of
dots and crosses:

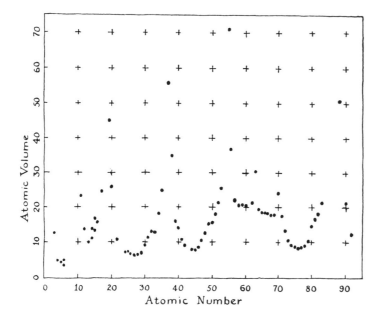

The space opened up by erasing can be effectively used. Labels for the initial elements of each period, an alkali, show the beginning of each cycle in the periodic table of elements—and in the graphic. The unusual rare-earths are indicated. In addition, the label and numbers on the vertical axis are turned to read from left to right rather than bottom to top, making the graphic slightly more accessible, a little more friendly:

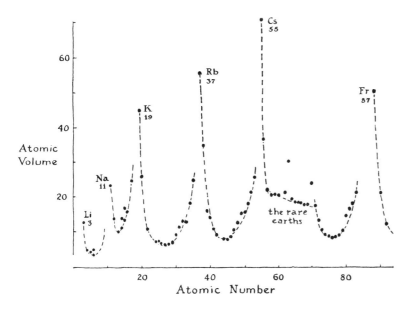

Conclusion

Five principles in the theory of data graphics produce substantial changes in graphical design. The principles apply to many graphics and yield a series of design options through cycles of graphical revision and editing.

Above all else show the data.

Maximize the data–ink ratio.

Erase non–data-ink.

Erase redundant data-ink.

Revise and edit.

With savage pictures fill their gaps
And o'er unhabitable downs
Place elephants for want of towns.

Jonathan Swift's indictment of 17th-century cartographers

5   Chartjunk: Vibrations, Grids, and Ducks

The interior decoration of graphics generates a lot of ink that does not tell the viewer anything new. The purpose of decoration varies —to make the graphic appear more scientific and precise, to enliven the display, to give the designer an opportunity to exercise artistic skills. Regardless of its cause, it is all non-data-ink or redundant data-ink, and it is often chartjunk. Graphical decoration, which prospers in technical publications as well as in commercial and media graphics, comes cheaper than the hard work required to produce intriguing numbers and secure evidence.

Sometimes the decoration is thought to reflect the artist's fundamental design contribution, capturing the essential spirit of the data and so on. Thus principles of artistic integrity and creativity are invoked to defend—even to advance—the cause of chartjunk. There are better ways to portray spirits and essences than to get them all tangled up with statistical graphics.

Fortunately most chartjunk does not involve artistic considerations. It is simply conventional graphical paraphernalia routinely added to every display that passes by: over-busy grid lines and excess ticks, redundant representations of the simplest data, the debris of computer plotting, and many of the devices generating design variation.

Like weeds, many varieties of chartjunk flourish. Here three widespread types found in scientific and technical research work are catalogued—unintentional optical art, the dreaded grid, and the self-promoting graphical duck. A hundred chartjunky examples from commercial and media graphics have been forgone so as to demonstrate the relevance of the critique to the professional scientific production of data graphics.

Unintentional Optical Art

Contemporary optical art relies on moiré effects, in which the design interacts with the physiological tremor of the eye to produce the distracting appearance of vibration and movement.

The effect extends beyond the ink of the design to the whole page. When exploited by the experts, such as Bridget Riley and Victor Vasarely, op art effects are undoubtedly eye-catching.

But statistical graphics are also often drawn up so as to shimmer. This moiré vibration, probably the most common form of graphical clutter, is inevitably bad art and bad data graphics. The noise clouds the flow of information as these examples from technical and scientific publications illustrate:

Instituto de Expansão Commercial, *Brasil: Graphicos Economicos–Estatisticas* (Rio de Janeiro, 1929), p. 15.

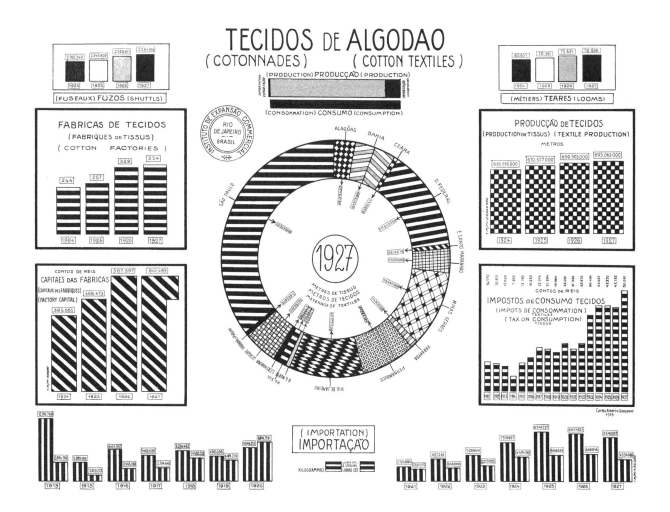

Year-End Primary Stocks of Crude Oil and Refined Petroleum Products

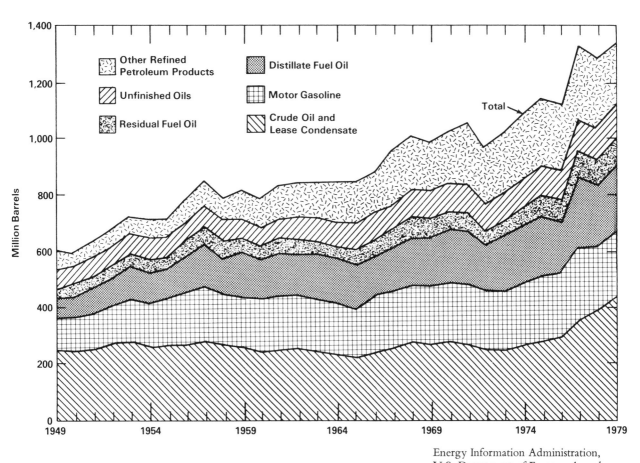

Energy Information Administration, U.S. Department of Energy, *Annual Report to Congress, 1979* (Washington, D.C., 1979), vol. 2, p. 64.

Moiré vibration appears at a maximum for equally spaced bars:

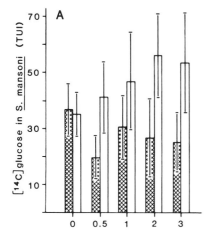

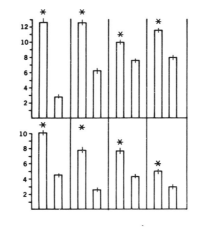

James T. Kuznicki and N. Bruce McCutcheon, "Cross-Enhancement of the Sour Taste on Single Human Taste Papillae," *Journal of Experimental Psychology: General*, 108 (1979), 76.

Eain M. Cornford and Marie E. Huot, "Glucose Transfer from Male to Female Schistosomes," *Science*, 213 (September 11, 1981), 1270.

And, finally, from the style sheet once provided by the *Journal of the American Statistical Association*, a graphic described as "an example of a figure prepared in the proper form":

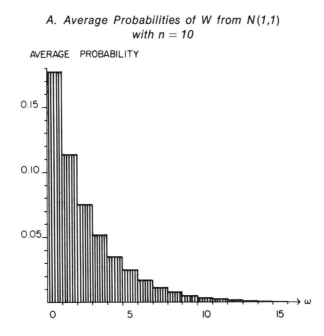

"JASA Style Sheet," *Journal of the American Statistical Association*, 71 (March 1976), 260–261.

The display required 131 line-strokes and 15 digits to communicate its simple information. The vibrating lines are poorly drawn, unevenly spaced, and misaligned with the vertical axis.

Vibrating chartjunk even frequents the graphics of major scientific journals:

The ten most frequently cited (footnoted) scientific journals: random sample of issues published 1980–1982	Percentage of graphics with moiré vibration	Number of graphics in sample
Biochemistry	2%	568
Journal of Biological Chemistry	2%	565
Journal of the American Chemical Society	3%	317
Journal of Chemical Physics	6%	327
Biochimia et Biophysica Acta	8%	432
Nature	11%	225
Proceedings of the National Academy of Sciences, U.S.A.	12%	438
Lancet	15%	364
Science	17%	311
New England Journal of Medicine	21%	338

Moiré effects have proliferated in computer graphics and with
the widespread use of transfer designs printed on thin plastic sheets.
The transfer sheets are filled with instant chartjunk. Shown here
are a few of the vibrating possibilities from a catalog. Cross-
hatching should be replaced with screens of varying density
or shades of gray. Specific areas on the graphic should be labeled
with words rather than encoded with different types of hatching.

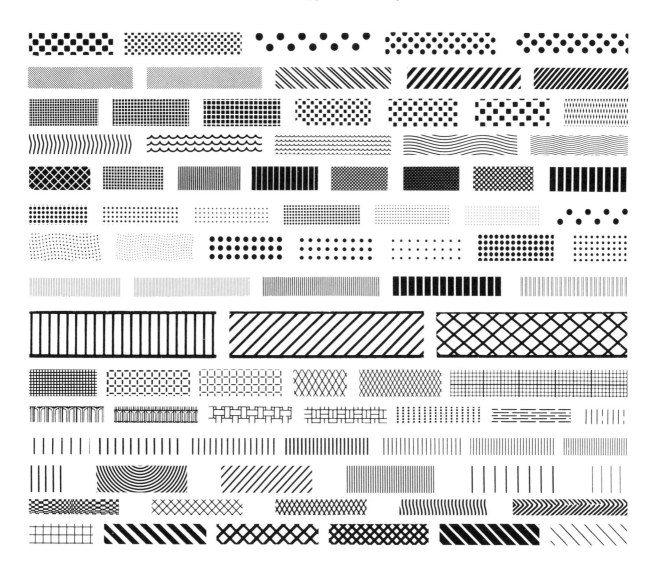

This form of chartjunk is a twentieth-century innovation, and
computer graphics are multiplying it more than ever. The handbooks
and textbooks of statistical graphics, along with user's manuals
for computer graphics programs, are filled up with vibrating
graphics, presented as exemplars of design. Note the high

proportion of chartjunky graphics in the more recent publications. Computer graphics are particularly active:

Textbooks and handbooks of statistical graphics; and manuals for computer graphics programs (ordered by date of publication)	Percentage of graphics with moiré vibration	Total number of graphics
William C. Brinton, *Graphic Methods for Presenting Facts* (New York, 1914)	12%	255
R. Satet, *Les Graphiques* (Paris, 1932)	29%	28
Herbert Arkin and Raymond R. Colton, *Graphs: How to Make and Use Them* (New York, 1936)	17%	95
Mary Eleanor Spear, *Charting Statistics* (New York, 1952)	46%	134
Anna C. Rogers, *Graphic Charts Handbook* (Washington, D.C., 1961)	32%	201
F. J. Monkhouse and H. R. Wilkinson, *Maps and Diagrams* (London, third edition, 1971)	14%	322
Calvin F. Schmid and Stanton E. Schmid, *Handbook of Graphic Presentation* (New York, second edition, 1979)	22%	399
A. J. MacGregor, *Graphics Simplified* (Toronto, 1979)	34%	65
The user's manual for a widely distributed computer graphics package: *SAS/GRAPH User's Guide* (Cary, North Carolina, 1980)	68%	28
The manual for a very extensive computer graphics program: *Tell-A-Graf User's Manual* (San Diego, 1981)	53%	459

Can optical art effects ever produce a better graphic? Bertin exhorts: "It is the designer's duty to make the most of this variation; to obtain the resonance [of moiré vibration] without provoking an uncomfortable sensation: to flirt with ambiguity without succumbing to it."[1] But can statistical graphics successfully "flirt with ambiguity"? It is a clever idea, but no good examples are to be found. The key difficulty remains: moiré vibration is an *undisciplined* ambiguity, with an illusive, eye-straining quality that contaminates the entire graphic. It has no place in data graphical design.

[1]Jacques Bertin, *Semiologie Graphique* (Paris, second edition, 1973), p. 80, translated by William J. Berg and Howard Wainer for forthcoming English edition.

The Grid

One of the more sedate graphical elements, the grid should usually be muted or completely suppressed so that its presence is only implicit—lest it compete with the data. Grids are mostly for the initial plotting of data at home or office rather than for putting into print. Dark grid lines are chartjunk. They carry no information, clutter up the graphic, and generate graphic activity unrelated to data information. This grid camouflages the profile of the data in the age-sex pyramid of the population of France in 1968:

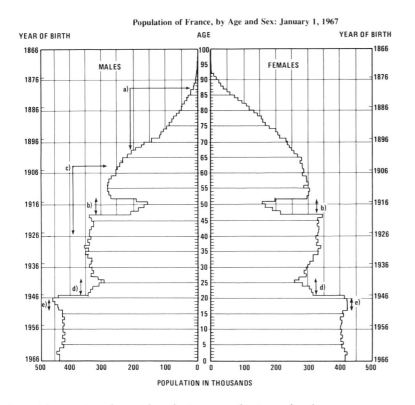

Population of France, by Age and Sex: January 1, 1967

(a) Military losses in World War I
(b) Deficit of births during World War I
(c) Military losses in World War II
(d) Deficit of births during World War II
(e) Rise of births due to demobilization after World War II

A revision quiets the grid and gives emphasis to the data:

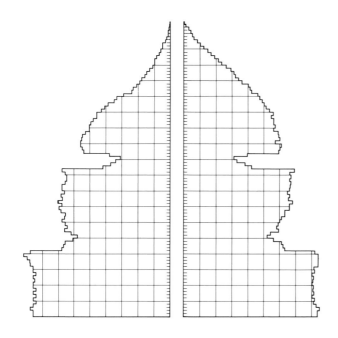

Based on data in Institut National de la Statistique et des Études Économiques, *Annuaire statistique de la France, 1968* (Paris, 1968), pp. 32–33; redrawn in Henry S. Shryock and Jacob S. Siegel, *The Methods and Materials of Demography* (Washington, D.C., 1973), vol. 1, 242.

The space occupied by the doubled grid lines consumes 18 percent of the area of this otherwise most ingenious design, a "multi-window plot." Optical white dots appear at the intersections of the grid lines. (The plot shows the following: The large square contains X_4, X_7 scatterplots for the indicated levels of X_1 and X_3. The marginal plots on the right are conditioned on X_3 and the plots at the top on X_1. The upper right corner shows the unconditional X_4, X_7 scatter.) Redrawing eliminates the vibration:

Paul A. Tukey and John W. Tukey, "Data-Driven View Selection; Agglomeration and Sharpening," in Vic Barnett, ed., *Interpreting Multivariate Data* (Chichester, England, 1981), 231–232.

MULTIWINDOW PLOT OF PARTICLE PHYSICS MOMENTUM DATA

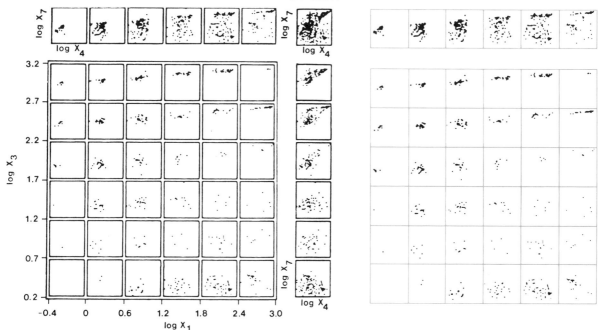

The grid in the classic Marey train schedule is very active:

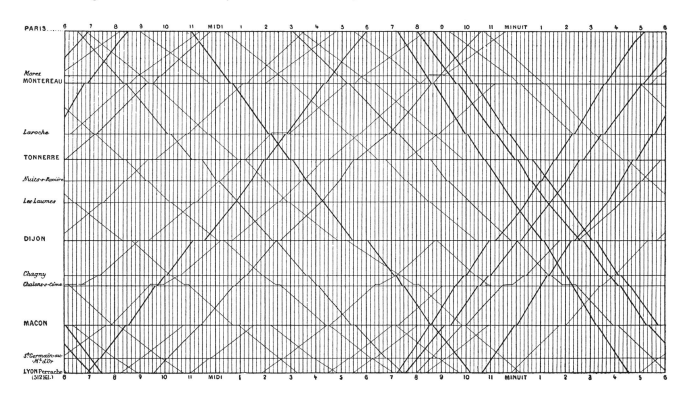

Thinning the grid lines helps a little bit:

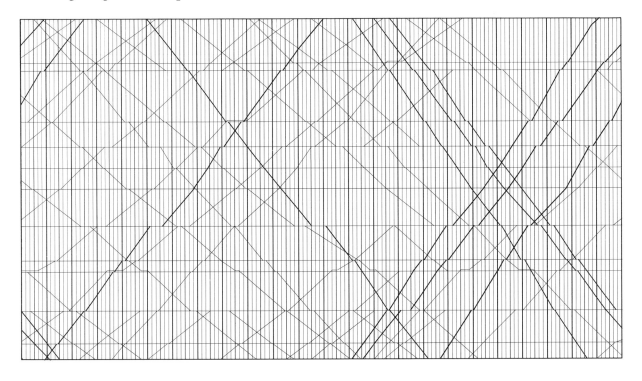

A better treatment, however, is a *gray grid*:

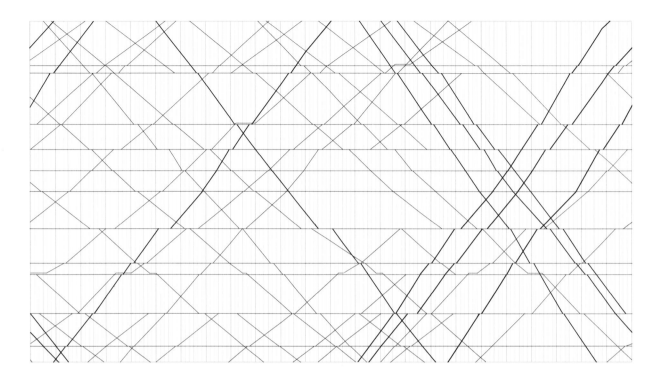

When a graphic serves as a look-up table, then a grid may help in reading and interpolating. But even in this case the grids should be muted relative to the data. A gray grid works well and, with a delicate line, may promote more accurate data reconstruction than a dark grid.

Most ready-made graph paper comes with a darkly printed grid. The reverse (unprinted) side should be used, for then the lines show through faintly and do not clutter the data. If the paper is heavily gridded on both sides, throw it out.

Self-Promoting Graphics: The Duck

When a graphic is taken over by decorative forms or computer debris, when the data measures and structures become Design Elements, when the overall design purveys Graphical Style rather than quantitative information, then that graphic may be called a *duck* in honor of the duck-form store, "Big Duck." For this building the whole structure is itself decoration, just as in the duck data graphic. In *Learning from Las Vegas*, Robert Venturi, Denise Scott

Brown, and Steven Izenour write about the ducks of modern architecture—and their thoughts are relevant to the design of data graphics as well:

> When Modern architects righteously abandoned ornament on buildings, they unconsciously designed buildings that *were* ornament. In promoting Space and Articulation over symbolism and ornament, they distorted the whole building into a duck. They substituted for the innocent and inexpensive practice of applied decoration on a conventional shed the rather cynical and expensive distortion of program and structure to promote a duck. . . . It is now time to reevaluate the once-horrifying statement of John Ruskin that architecture is the decoration of construction, but we should append the warning of Pugin: It is all right to decorate construction but never construct decoration.[2]

[2] Robert Venturi, Denise Scott Brown, and Steven Izenour, *Learning from Las Vegas* (Cambridge, revised edition, 1977), p. 163. The initial statement of the duck concept is found on pp. 87–103.

Peter Blake, *God's Own Junkyard* (New York, 1964, 1979), p. 101.

The addition of a fake perspective to the data structure clutters many graphics. This variety of chartjunk, now at high fashion in the world of Boutique Data Graphics, abounds in corporate annual reports, the phony statistical studies presented in advertisements, the mass media, and the more muddled sorts of social science research.

A series of weird three-dimensional displays appearing in the magazine *American Education* in the 1970s delighted connoisseurs of the graphically preposterous. Here five colors report, almost by happenstance, only five pieces of data (since the division within each year adds to 100 percent). This may well be the worst graphic ever to find its way into print:

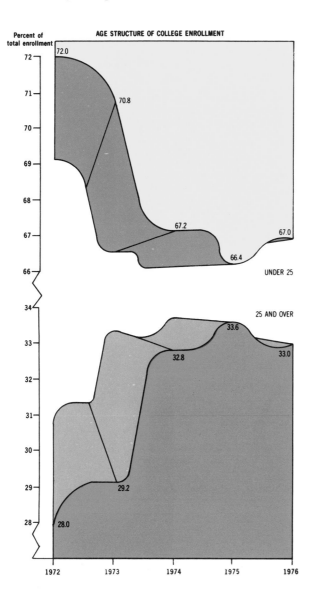

There are some superbly produced ducks:

William L. Kahrl, et al., *The California Water Atlas* (Sacramento, 1978, 1979), p. 55.

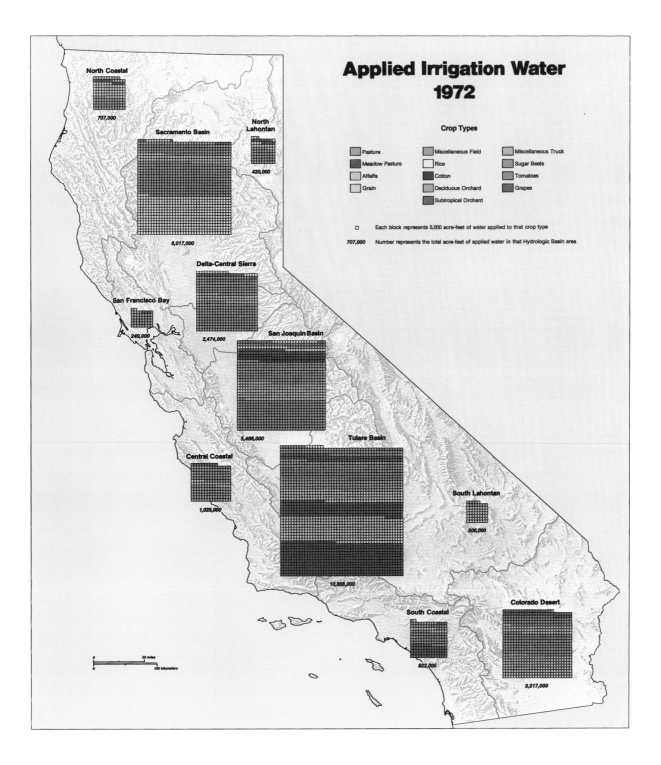

Occasionally designers seem to seek credit merely for possessing a new technology, rather than using it to make better designs. Computers and their affiliated apparatus can do powerful things graphically, in part by turning out the hundreds of plots necessary for good data analysis. But at least a few computer graphics only evoke the response "Isn't it remarkable that the computer can be programmed to draw like that?" instead of "My, what interesting data."

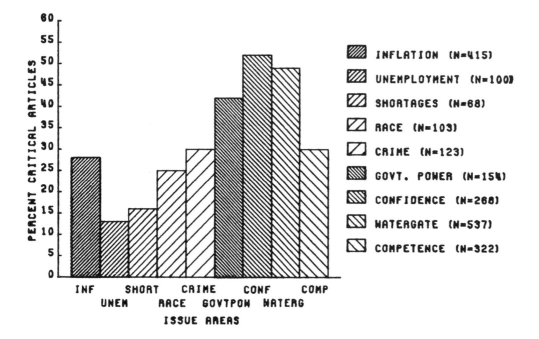

The symptoms of the We-Used-A-Computer-To-Build-A-Duck Syndrome appear in this display from a professional journal: the thin substance; the clotted, crinkly lettering all in upper-case sans serif; the pointlessly ordered cross-hatching; the labels written in computer abbreviations; the optical vibration—all these the by-products of the technology of graphic fabrication. The overly busy vertical scaling shows more percentage markers and labels than there are actual data points. The observed values of the percentages should be printed instead. Since the information consists of a few numbers and a good many words, it is best to pass up the computerized graphics capability this time and tell the story with a table:

Arthur H. Miller, Edie N. Goldenberg, and Lutz Erbring, "Type-Set Politics: Impact of Newspapers on Public Confidence," *American Political Science Review*, 73 (1979), 67–84.

Content and tone of front-page articles in 94 U.S. newspapers, October and November, 1974	Number of articles	Percent of articles with negative criticism of specific person or policy
Watergate: defendants and prosecutors, Ford's pardon of Nixon	537	49%
Inflation, high cost of living	415	28%
Government competence: costs, quality, salaries of public employees	322	30%
Confidence in government: power of special interests, trust in political leaders, dishonesty in politics	266	52%
Government power: regulation of business, secrecy, control of CIA and FBI	154	42%
Crime	123	30%
Race	103	25%
Unemployment	100	13%
Shortages: energy, food	68	16%

Conclusion

Chartjunk does not achieve the goals of its propagators. The overwhelming fact of data graphics is that they stand or fall on their content, gracefully displayed. Graphics do not become attractive and interesting through the addition of ornamental hatching and false perspective to a few bars. Chartjunk can turn bores into disasters, but it can never rescue a thin data set. The best designs (for example, Minard on Napoleon in Russia, Marey's graphical train schedule, the cancer maps, the *Times* weather history of New York City, the chronicle of the annual adventures of the Japanese beetle, the new view of the galaxies) are *intriguing and curiosity-provoking*, drawing the viewer into the wonder of the data, sometimes by narrative power, sometimes by immense detail, and sometimes by elegant presentation of simple but interesting data. But no information, no sense of discovery, no wonder, no substance is generated by chartjunk.

Forgo chartjunk, including

moiré vibration,

the grid, and the duck.

Painting is special, separate, a matter of meditation and contemplation,
for me, no physical action or social sport. As much consciousness as
possible. Clarity, completeness, quintessence, quiet. No noise, no schmutz,
no schmerz, no fauve schwärmerei. Perfection, passiveness, consonance,
consummateness. No palpitations, no gesticulation, no grotesquerie.
Spirituality, serenity, absoluteness, coherence. No automatism,
no accident, no anxiety, no catharsis, no chance. Detachment,
disinterestedness, thoughtfulness, transcendence. No humbugging,
no button-holing, no exploitation, no mixing things up.

Ad Reinhardt, statement for the catalogue of the exhibition, "The New Decade: 35 American Painters and Sculptors," Whitney Museum of American Art, New York, 1955.

6 *Data-Ink Maximization and Graphical Design*

So far the principles of maximizing data-ink and erasing have helped to generate a series of choices in the process of graphical revision. This is an important result, but can the ideas reach beyond the details and particularities of editing? Is it possible to do what a theory of graphics is supposed to do, that is, to derive new graphical forms? In this chapter the principles are applied to many graphical designs, basic and advanced, including box plots, bar charts, histograms, and scatterplots. New designs result.

Redesign of the Box Plot

Mary Eleanor Spear's "range bar"

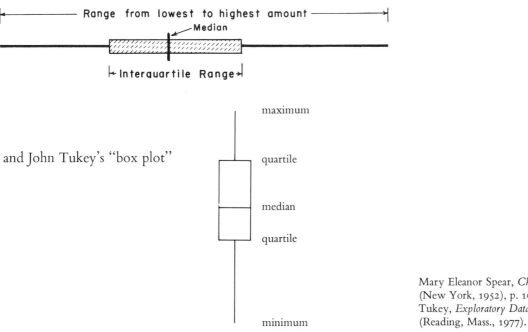

and John Tukey's "box plot"

Mary Eleanor Spear, *Charting Statistics* (New York, 1952), p. 166; and John W. Tukey, *Exploratory Data Analysis* (Reading, Mass., 1977).

can be mostly erased without loss of information:

The revised design, a *quartile plot*, shows the same five numbers. It is easy to draw by hand or computer and, most importantly, can replace the conventional scatterplot frame. The straightedge need only be placed on the paper once to draw the quartile plot, compared to six separate placings for the box plot. An alternative is

but this design will not work effectively to frame a scatterplot. Nor does it look very good.

Perhaps special emphasis should be given to the middle half of the distribution, however, as in the box plot. This can be done by changing line weights

or, even better, by offsetting the middle half:

This latter design is the preferred form of the quartile plot. It uses the ink effectively and looks good.

In these revisions of the box plot, the principle of maximizing data-ink has suggested a variety of designs, but the choice of the best overall arrangement naturally also rests on statistical and aesthetic criteria—in other words, the procedure is one of *reasonable* data-ink maximizing.

The same logic applies to many similar designs, such as this "parallel schematic plot." The original required 80 separate placings of the straightedge, 50 horizontals and 30 verticals:

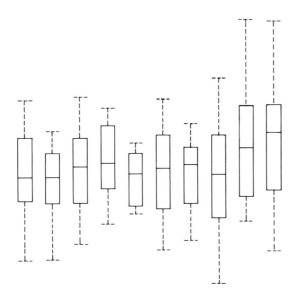

An erased version requires only 10 verticals to show the same information:

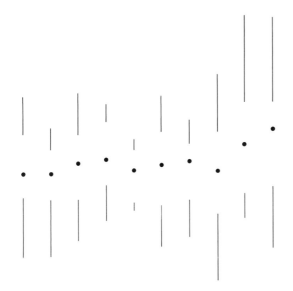

The large reduction in the amount of drawing is relevant for the use of such designs in informal, exploratory data analysis, where the research worker's time should be devoted to matters other than drawing lines.

Redesign of the Bar Chart/Histogram

Here is the standard model bar chart, with the design endorsed by
the practices and the style sheets of many statistical and scientific
publications:

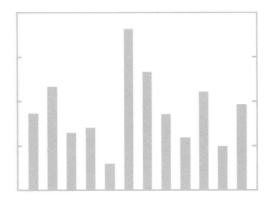

Its architecture differs little from Playfair's original design:

Exports and Imports of SCOTLAND to and from different parts for one Year from Christmas 1780 to Christmas 1781

| 10 20 30 40 50 60 70 80 90 100 110 | 130 | 150 | 170 | 200 | 200 | 240 | 260 | 280 | L.300,000 |

Names of Places.

Jersey &c.
Ireland
Poland
Isle of Man
Greenland
Prussia
Portugal
Holland
Sweden
Guernsey
Germany
Denmark and Norway
Flanders
West Indies
America
Russia
Ireland.

The Upright divisions are Ten Thousand Pounds each.   The Black Lines are Exports the Ribbed lines Imports

Published as the Act directs June 7th 1786 by Wm Playfair   Neele sculp 352 Strand. London.

The box can be erased:

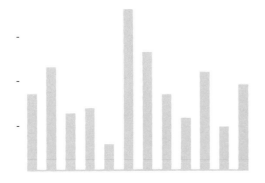

And the vertical axis, except for the ticks:

Even part of the data measures can be erased, making a *white grid*, which shows the coordinate lines more precisely than ticks alone:

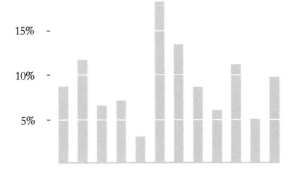

The white grid eliminates the tick marks, since the numerical labels
on the vertical are tied directly to the white lines:

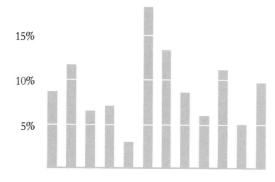

Although the intersection of the thicker bar with the thinner baseline creates an attractive visual effect (but also the optical illusion
of gray dots at the intersections), the baseline can be erased since
the bars define the end-point at the bottom:

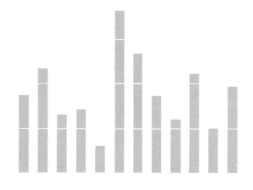

Still, a thin baseline looks good:

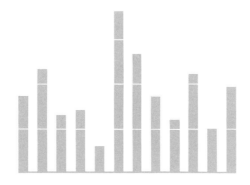

Erasing and data-ink maximizing have induced changes in the plain old bar chart. The techniques—no frame, no vertical axis, no ticks, and the white grid—apply to other designs:

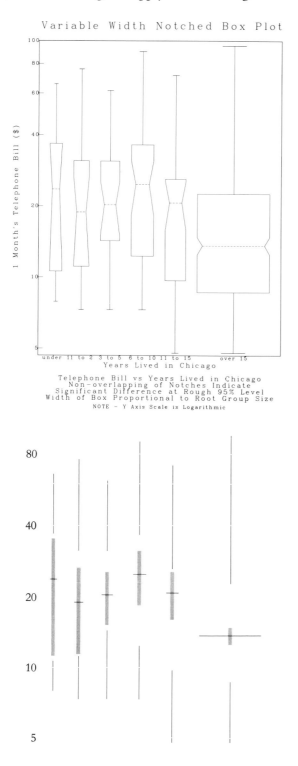

Variable Width Notched Box Plot

Telephone Bill vs Years Lived in Chicago
Non-overlapping of Notches Indicate
Significant Difference at Rough 95% Level
Width of Box Proportional to Root Group Size
NOTE – Y Axis Scale is Logarithmic

Robert McGill, John W. Tukey, and Wayne A. Larsen, "Variations of Box Plots," *American Statistician*, 32 (1978), 12–16.

Redesign of the Scatterplot

Consider the standard bivariate scatterplot:

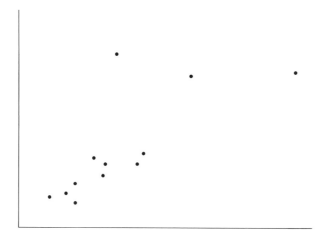

A useful fact, brought to notice by the maximization and erasing principles, is that the frame of a graphic can become an effective data-communicating element simply by erasing part of it. The frame lines should extend only to the measured limits of the data rather than, as is customary, to some arbitrary point like the next round number marking off the grid and grid ticks of the plot. That part of the frame exceeding the limits of the observed data is trimmed off:

The result, a *range-frame*, explicitly shows the maximum and minimum of both variables plotted (along with the range), information available only by extrapolation and visual estimation in the conventional design. The data-ink ratio has increased: some non-data-ink has been erased, and the remainder of the frame, now carrying information, has gone over to the side of data-ink.

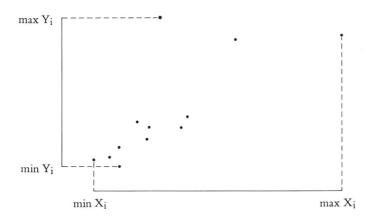

Nothing but the tails of the frame need change:

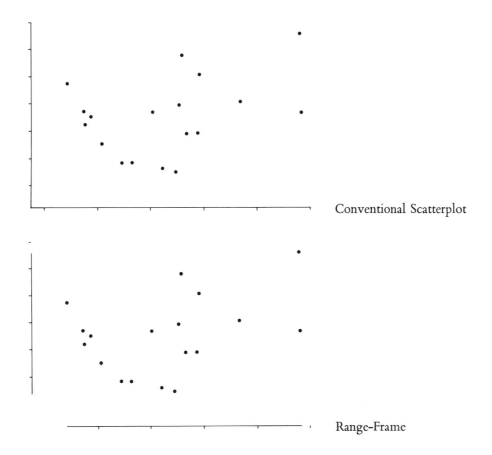

Conventional Scatterplot

Range-Frame

A range-frame does not require any viewing or decoding instructions; it is not a graphical puzzle and most viewers can easily tell what is going on. Since it is more informative about the data in a clear and precise manner, the range-frame should replace the non-data-bearing frame in many graphical applications.

A small shift in the remaining ink turns each range-frame into
a quartile plot:

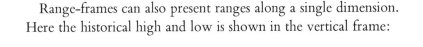

Erasing and editing has led to the display of ten extra numbers
(the minimum, maximum, two quartiles, and the median for both
variables). The design is useful for analytical and exploratory data
analysis, as well as for published graphics where summary char-
acterizations of the marginal distributions have interest. The design
is nearly always better than the conventionally framed scatterplot.

Range-frames can also present ranges along a single dimension.
Here the historical high and low is shown in the vertical frame:

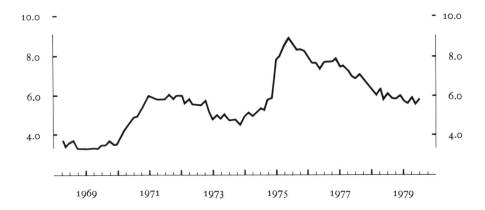

Finally, the entire frame can be turned into data by framing the bivariate scatter with the marginal distribution of each variable. The *dot-dash-plot* results.[1]

[1] The terminology follows tradition, for scatterplots were once called "dot diagrams"—for example, in R. A. Fisher's *Statistical Methods for Research Workers* (Edinburgh, 1925).

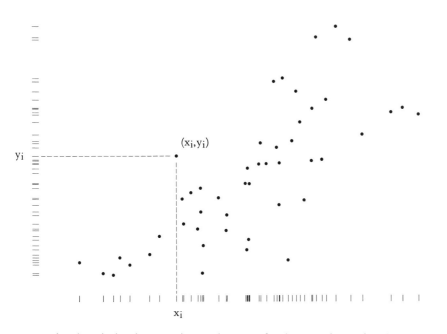

The dot-dash-plot combines the two fundamental graphical designs used in statistical analysis, the marginal frequency distribution and the bivariate distribution. Dot-dash-plots make routine what good data analysts do already—plotting marginal and joint distributions together.

An empirical cumulative distribution of residuals on a normal grid shows the outer 16 terms plus the 30th term, with all 60 points plotted in the marginal distribution:

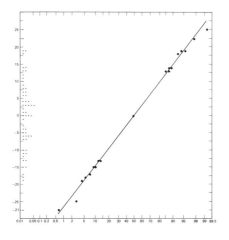

Cuthbert Daniel, *Applications of Statistics to Industrial Experimentation* (New York, 1976), p. 155.

Similarly, this data-rich graphic of signals from pulsars shows both marginal distributions:

Timothy H. Hankins and Barney J. Rickett, "Pulsar Signal Processing," in Berni Alder, et al., eds., *Methods in Computational Physics, Volume 14: Radio Astronomy* (New York, 1975), p. 108.

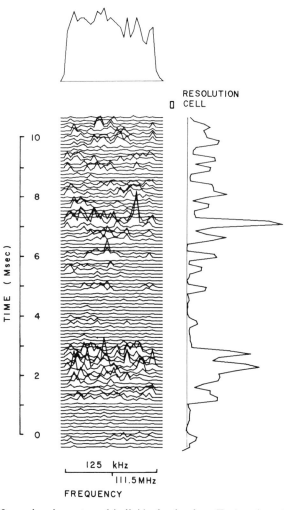

Narrowband spectra of individual subpulses. Each point of the intensity $I_q(t)$ plotted on the right is the sum of the distribution of intensities across the receiver bandwidth shown in the center. At the top is plotted the spectrum averaged over the pulse. In the limit of many thousands of pulses this would show the receiver bandpass shape.

The fringe of dashes in the dot-dash-plot can connect a series of bivariate scatters in a *rugplot* (since it resembles a set of fringed rugs—and covers the statistical ground):

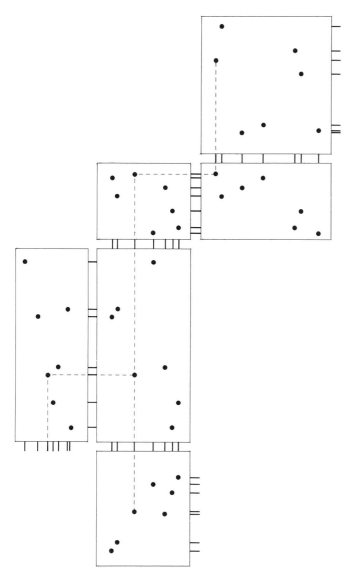

Reflecting the one-dimensional projections from each scatter, the dashes encourage the eye to notice how each plot filters and translates the data through the scatter from one adjacent plot to the next. Sometimes it is useful to think of each bivariate scatter as the imperfect empirical representation of an underlying curve that transforms one variable into another. In the rugplot, the sequence of variables can wander off as appropriate. The quantitative history of a single observation can be traced through a series of one- and two-dimensional contexts.

Conclusion

The first part of a theory of data graphics is in place. The idea, as described in the previous three chapters, is that most of a graphic's ink should vary in response to data variation. The theory has something to say about a great variety of graphics—workaday scientific charts, the unique drawings of Roger Hayward, the exemplars of graphical handbooks, newspaper displays, computer graphics, standard statistical graphics, and the recent inventions of Chernoff and Tukey.

The observed increases in efficiency, in how much of the graphic's ink carries information, are sometimes quite large. In several cases, the data-ink ratio increased from .1 or .2 to nearly 1.0. The transformed designs are less cluttered and can be shrunk down more readily than the originals.

But, are the transformed designs *better*?

(1) They are necessarily better within the principles of the theory, for more information per unit of space and per unit of ink is displayed. And this is significant; indeed, the history of devices for communicating information is written in terms of increases in efficiency of communication and production.

(2) Graphics are almost always going to improve as they go through editing, revision, and testing against different design options. The principles of maximizing data-ink and erasing generate graphical alternatives and also suggest a direction in which revisions should move.

(3) Then there is the audience: will those looking at the new designs be confused? Some of the designs are self-explanatory, as in the case of the range-frame. The dot-dash-plot is more difficult, although it still shows all the standard information found in the scatterplot. Nothing is lost to those puzzled by the frame of dashes, and something is gained by those who do understand. Moreover, it is a frequent mistake in thinking about statistical graphics to underestimate the audience. Instead, why not assume that if you understand it, most other readers will, too? Graphics should be as intelligent and sophisticated as the accompanying text.

(4) Some of the new designs may appear odd, but this is probably because we have not seen them before. The conventional designs for statistical graphics have been viewed thousands of times by nearly every reader of this book; on the other hand, the range-frame, the dot-dash-plot, the white grid, the quartile plot, the rugplot, and the half-face just a few times. With use, the new designs will come to look just as reasonable as the old.

Maximizing data ink (within reason) is but a single dimension of
a complex and multivariate design task. The principle helps conduct experiments in graphical design. Some of those experiments
will succeed. There remain, however, many other considerations
in the design of statistical graphics—not only of efficiency,
but also of complexity, structure, density, and even beauty.

7 *Multifunctioning Graphical Elements*

The same ink should often serve more than one graphical purpose. A graphical element may carry data information and also perform a design function usually left to non-data-ink. Or it might show several different pieces of data. Such *multifunctioning graphical elements*, if designed with care and subtlety, can effectively display complex, multivariate data.[1]

Consider, for example, the multifunctioning blot of the blot map. It simultaneously locates the geographic unit on a two-dimensional surface, describes the shape of the geographic unit, and indicates the level of the variable displayed by color or intensity of shading. That is a great deal of information for a small patch of ink—and the different pieces of information are not confounded and mixed together.

In contrast, the conventional graphical frame performs only a modest design function, the separation of the grid and data measures from the labels. And it is a place to hang the grid ticks. With all that ink doing so little, it is a prime candidate for mobilization as a double-functioning graphical element. Hence the range-frame, the quartile frame, and the dot-dash-plot.

The principle, then, is:

> Mobilize every graphical element, perhaps
> several times over, to show the data.

The danger of multifunctioning elements is that they tend to generate graphical puzzles, with encodings that can only be broken by their inventor. Thus design techniques for enhancing graphical clarity in the face of complexity must be developed along with multifunctioning elements.

Data-Built Data Measures

The graphical element that actually locates or plots the data is the *data measure*. The bars of a bar chart, the dots of a scatterplot, the dots and dashes of a dot-dash-plot, the blots of a blot map are data measures. The ink of the data measure can itself carry data; for example, the dots of the scatterplot can take on different shadings in response to a third variable.

[1] The idea of double-functioning elements appears in architectural criticism; see Robert Venturi, *Complexity and Contradiction in Architecture* (New York, second edition, 1977), ch. 5. Venturi in turn cites Wylie Sypher, *Four Stages of Renaissance Style* (Garden City, N.Y., 1955).

Building data measures out of the data increases the quantitative detail and dimensionality of a graphic. The stem-and-leaf plot constructs the distribution of a variable with numbers themselves:

0 | 9 = 900 feet

0	98766562
1	97719630
2	6998776654444222211009850
3	876655412099551426
4	99988443319294333361107
5	97666666554422210097731
6	898665441077761065
7	98855431100652108073
8	653322122937
9	377655421000493
10	0984433165212
11	4963201631
12	45421164
13	47830
14	00
15	676
16	52
17	92
18	5
19	39730

Stem-and-leaf displays:
heights of 218 volcanoes, unit 100 feet.

19 | 3 = 19,300 feet

The idea of making every graphical element effective was behind the design of the stem-and-leaf plot. In presenting his invention, John Tukey wrote: "If we are going to make a mark, it may as well be a meaningful one. The simplest—and most useful—meaningful mark is a digit."[2]

Here, too, the data form the data measure. Note the bimodal distribution in the histogram of college students arranged by height.

[2] "Some Graphic and Semigraphic Displays," in T. A. Bancroft, ed., *Statistical Papers in Honor of George W. Snedecor* (Ames, Iowa, 1972), p. 296.

Brian L. Joiner, "Living Histograms," *International Statistical Review*, 43 (1975), 339–340.

A distinguished graphic that builds data measures out of data was designed by Colonel Leonard P. Ayres for his statistical history of World War I, a book with several notable graphics all done by typewriter and rule. Constructing the data measures out of each American division's name (a numerical designation) turns what might have been a routine time-series into an elegant display. (Note that the cumulative design depends on the fact that none of the divisions returned before October 1918.) The triple-functioning data measure shows: (1) the number of divisions in France for each month, June 1917 to October 1918; (2) what particular divisions were in France in each month; and (3) the duration of each division's presence in France.

Leonard P. Ayres, *The War with Germany* (Washington, D.C., 1919), p. 102.

```
                                                  8
                                                 38
                                                 31
                                              34 34
                                              86 86
                                              84 84
                                              87 87
                                           40 40 40
                                           39 39 39
                                           88 88 88
                                           81 81 81
                                            7  7  7
                                           85 85 85
                                        36 36 36 36
                                        91 91 91 91
                                        79 79 79 79
                                        76 76 76 76
                                     29 29 29 29 29
                                     37 37 37 37 37
                                     90 90 90 90 90
                                     92 92 92 92 92
                                     89 89 89 89 89
                                     83 83 83 83 83
                                     78 78 78 78 78
                                  80 80 80 80 80 80
                                  30 30 30 30 30 30
                                  33 33 33 33 33 33
                                   6  6  6  6  6  6
                                  27 27 27 27 27 27
                                   4  4  4  4  4  4
                                  28 28 28 28 28 28
                                  35 35 35 35 35 35
                                  82 82 82 82 82 82
                               77 77 77 77 77 77 77
                             3  3  3  3  3  3  3  3
                             5  5  5  5  5  5  5  5
                          32 32 32 32 32 32 32 32 32
                       41 41 41 41 41 41 41 41 41 41
                    42 42 42 42 42 42 42 42 42 42 42
       26 26 26 26 26 26 26 26 26 26 26 26 26 26 26
    2  2  2  2  2  2  2  2  2  2  2  2  2  2  2  2
 1  1  1  1  1  1  1  1  1  1  1  1  1  1  1  1  1
Jun Jul Aug Sep Oct Nov Dec | Jan Feb Mar Apr May Jun Jul Aug Sep Oct
           1917                         1918
```

Encoding of data measures can be far more elaborate. The plotted points here are Chernoff faces, which reduce well, maintaining legibility even with individual areas of .05 square inches as shown.[3] The analyst would observe the standard X-Y scatterplot and then turn to the within-scatter detail, seeking clusters of similar observations over the X-Y plane. Outlying faces and those inconsistent with others in the neighborhood—they are, of course, *strangers*—should be identified by observation number or name.

[3] Herman Chernoff, "The Use of Faces to Represent Points in k-Dimensional Space Graphically," *Journal of the American Statistical Association* 68 (June 1973), 361–368. For an application of faces located over two dimensions, see Howard Wainer and David Thissen, "Graphical Data Analysis," *Annual Review of Psychology*, 32 (1981), 191–241.

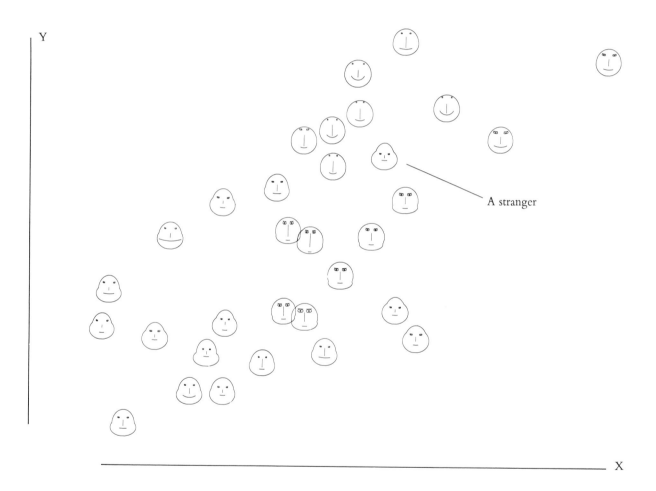

A stranger

With cartoon faces and even numbers becoming data measures, we would appear to have reached the limit of graphical economy of presentation, imagination, and, let it be admitted, eccentricity.

But let us consider this shaped poem, "Easter Wings" by George Herbert (1593–1633), which uses space—the length of each line— to depict quantity, all done 150 years before Playfair. The lines double-function: the longer lines describe wealth, plenty, largesse, and rising to flight; shorter lines tell of poverty and becoming "most thinne"; and lines of intermediate length indicate transition and change (decaying, rising, combining, becoming):

Easter-wings.

LOrd, who createdst man in wealth and store,
Though foolishly he lost the same,
Decaying more and more,
Till he became
Most poore:
With thee
O let me rise
As larks, harmoniously,
And sing this day thy victories:
Then shall the fall further the flight in me.

My tender age in sorrow did beginne:
And still with sicknesses and shame
Thou didst so punish sinne,
That I became
Most thinne.
With thee
Let me combine
And feel this day thy victorie:
For, if I imp my wing on thine,
Affliction shall advance the flight in me.

And the typographical delight of the statistician W. J. Youden:

THE
NORMAL
LAW OF ERROR
STANDS OUT IN THE
EXPERIENCE OF MANKIND
AS ONE OF THE BROADEST
GENERALIZATIONS OF NATURAL
PHILOSOPHY ◆ IT SERVES AS THE
GUIDING INSTRUMENT IN RESEARCHES
IN THE PHYSICAL AND SOCIAL SCIENCES AND
IN MEDICINE AGRICULTURE AND ENGINEERING ◆
IT IS AN INDISPENSABLE TOOL FOR THE ANALYSIS AND THE
INTERPRETATION OF THE BASIC DATA OBTAINED BY OBSERVATION AND EXPERIMENT

Finally, this graphical pun: the visual data as the data measure, as in the living histogram. The chart shows how states once differed in their engineering standards for painting lane stripes on road pavement. Some states marked the road lanes with short dashes and long gaps; others used only solid lines. Portrayed in the graphic is the actual physical pattern painted on the road, with 48 U.S. states ordered by the length of the painted mark:

Redrawn from A. R. Lauer, "Psychological Factors in Effective Traffic Control Devices," *Traffic Quarterly*, 5 (January 1951), 94.

Data-Based Grids

Very occasionally the grid can report directly on the data. This grid is formed by the location of measurement instruments; the plain dots register a zero reading, in contrast with the white background where no readings were taken. Erasing the grid would erase measured data (rather uneventful, to be sure). Such is not the case for most grid dots, ticks, and lines.

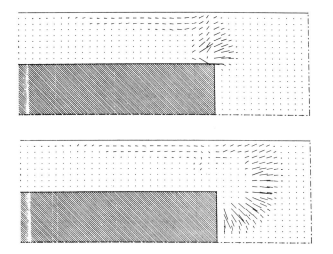

K. V. Roberts and D. E. Potter, "Magnetohydrodynamic Calculations," in Berni Alder, et al., eds., *Methods in Computational Physics: Volume 9, Plasma Physics* (New York, 1970), p. 402.

The arrangement of data in this table-graphic yields an internal grid, a rare example of data as grid:

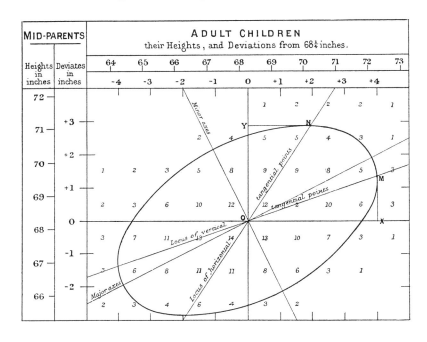

Karl Pearson, *The Life, Letters and Labours of Francis Galton* (Cambridge, 1930), vol. III-A, 14.

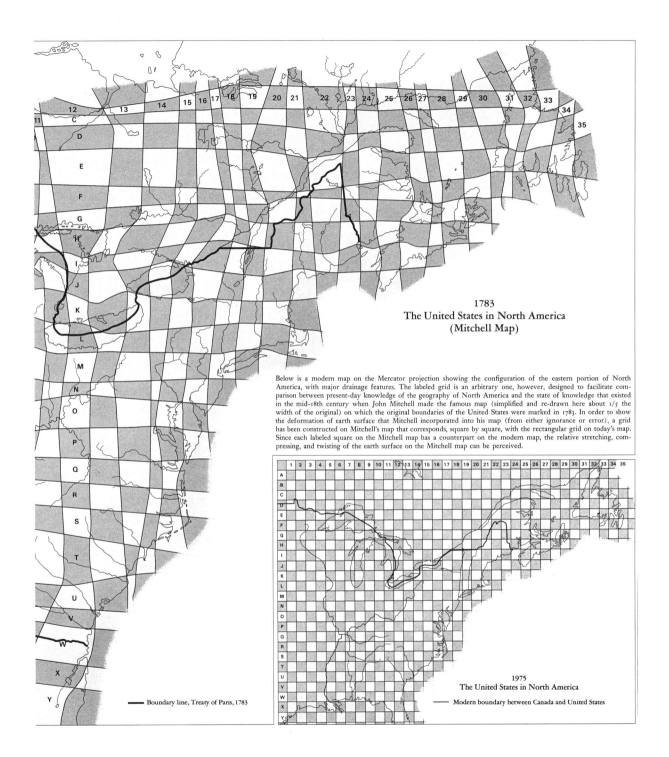

1783
The United States in North America
(Mitchell Map)

Below is a modern map on the Mercator projection showing the configuration of the eastern portion of North America, with major drainage features. The labeled grid is an arbitrary one, however, designed to facilitate comparison between present-day knowledge of the geography of North America and the state of knowledge that existed in the mid-18th century when John Mitchell made the famous map (simplified and re-drawn here about 1/5 the width of the original) on which the original boundaries of the United States were marked in 1783. In order to show the deformation of earth surface that Mitchell incorporated into his map (from either ignorance or error), a grid has been constructed on Mitchell's map that corresponds, square by square, with the rectangular grid on today's map. Since each labeled square on the Mitchell map has a counterpart on the modern map, the relative stretching, compressing, and twisting of the earth surface on the Mitchell map can be perceived.

—— Boundary line, Treaty of Paris, 1783

1975
The United States in North America

—— Modern boundary between Canada and United States

Here the grid is the element of interest, rather than the map.

Lester J. Cappon, Barbara Bartz Petchenik, and John Hamilton Long, *Atlas of Early American History* (Princeton, 1976), p. 58.

The grid that follows presents the data on the surface of the rock; on the sides, the grid is conventional. The two displays compare the effect of religion, taking into account party affiliation, on a person's vote for president in 1956 and in 1960 (when a Catholic ran for president). Note there is no reliable slope associated with religion in 1956, once party is controlled; in 1960, a systematic effect is found. Reading the slopes in the other direction shows the persistent effect of party in both elections:

Philip E. Converse, "Religion and Politics: The 1960 Election," in Angus Campbell, Philip E. Converse, Warren E. Miller, and Donald E. Stokes, *Elections and the Political Order* (New York, 1966), 102–103.

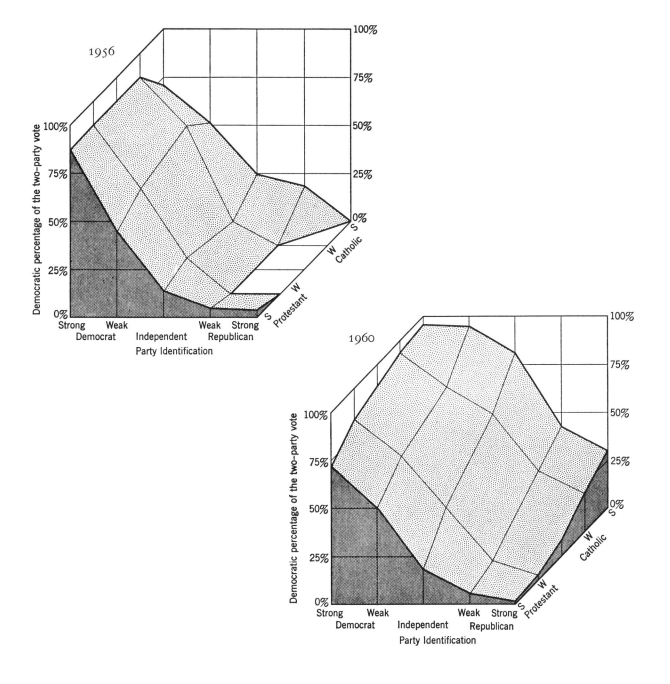

Playfair tied the grid to the data in his skyrocketing debt graphic. Although the implicit plotting coordinates are based on regular intervals, the vertical grid lines in the published version are irregularly spaced, keyed to significant events. The data-based grid is a shrewd graphical device, serving rather than fighting with the data. It is a technique underused in contemporary graphical work.

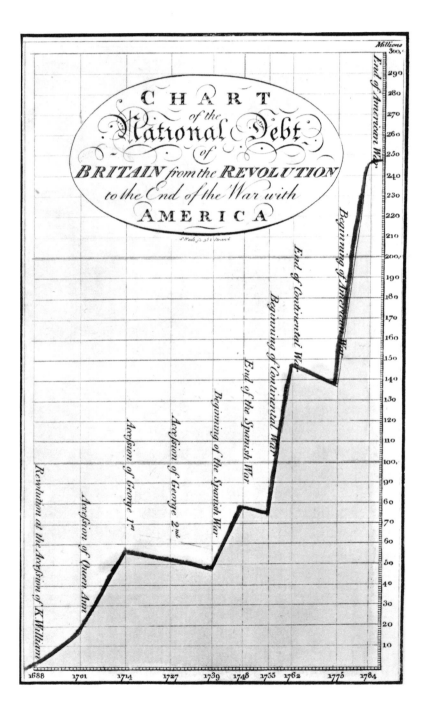

Double-Functioning Labels

Data-based coordinate lines lead to *data-based labels*, as, for example, at the bottom of Playfair's debt graphic. Again, the issue is the same: why not use the ink to show data? Beginning with conventionally labeled frame

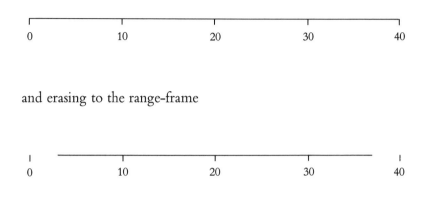

and erasing to the range-frame

leaves those lonely ticks and numbers out on the tails, working to help the eye get a better reading on where the line of the range-frame ends. But that job can be done better by investing the same ink in data: rather than showing the minimum round number and the maximum round number at the ends of the frame, show the actual minimum and maximum realized in the data:

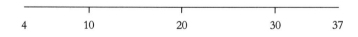

With its greater precision and two tick-marks less of non-data-ink, the range-frame with range-labels is superior to the range-frame with round number labels. Both improve on the standard, passive frame.

Numbers also double-function when used both to name things (like an identification number) and to reflect an ordering. In this graphic (in which the circled numbers fail to double-function), each number identifies a particular study of the thermal conductivity of tungsten, ordered alphabetically by the last name of the first author. If that list were ordered by date of publication instead, then the code would also indicate the time order in which the various conductivity determinations were made. Thus "1" would indicate the earliest study, and so on—or, alternatively, "61c" would be the third study published in 1961. Such information has interest, since we could see which of the early studies got the right answer. In addition, the movement of the studies toward the "correct" recommended values could be tracked. This extra information requires no additional ink.

C. Y. Ho, R. W. Powell, and P. E. Liley, *Thermal Conductivity of the Elements: A Comprehensive Review*, supplement no. 1, *Journal of Physical and Chemical Reference Data*, 3 (1974), I–692.

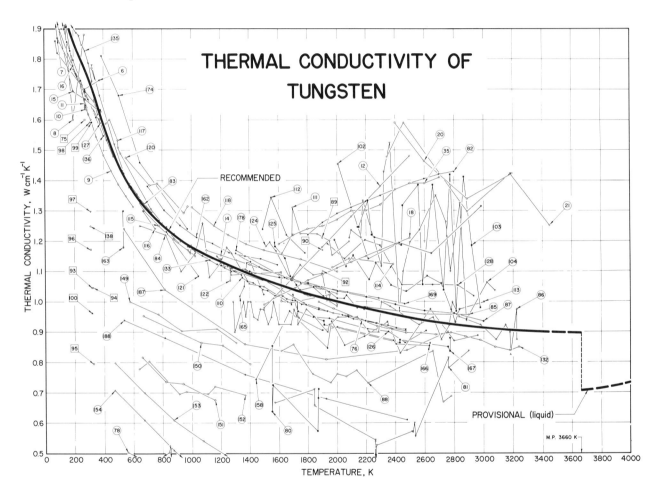

In most graphics, the coordinate labels are far from the data measures. Consequently the eye of the viewer must move back and forth between the path formed by the data and the coordinate positions arrayed along the margins of the graphic. Sometimes this eye-work can be eliminated entirely by turning the coordinate labels into data measures, another double-functioning maneuver. Take the example from the style sheet of the *Journal of the American Statistical Association*:

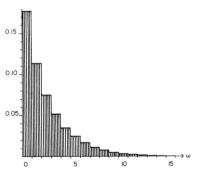

The grid increments of the X-axis are relocated upward to mark
the path of the data:

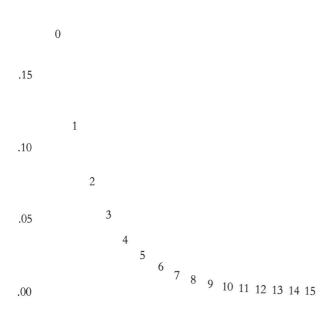

And since the issue in this display is the probability at each integer
value, the round-number Y-scale is replaced by exact values:

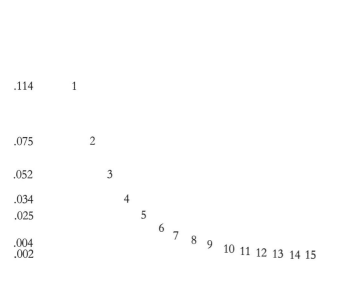

The Y-scale now resembles the dashes of the dot-dash-plot, with
the vertical column of data-positioned numbers serving as the
dashes to indicate the marginal distribution.

The method of data-based markers for the marginal distributions suggests a further enhancement of the dot-dash-plot:

20.3   •

15.2   •
14.6   •

11.3   •
10.1 •
8.4   •

5.1   •

     81   123   182   227 255 291   357

Now the numbers in the margin eliminate the standard frame and even a range-frame, replace the coordinate ticks, show the marginal distribution of both variables, and record the exact values of the two measurements made on each unit of observation. This graphical arrangement performs better for smaller data sets (say 30 observations or less) and when a fine level of detail is required.

Finally, a striking design with data-based labels:

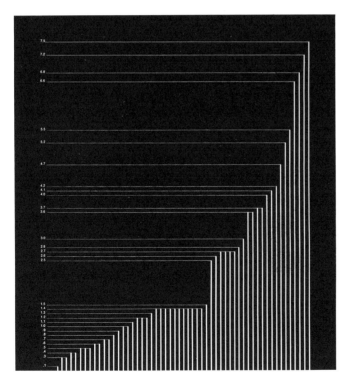

Designed by Carol Moore, Corporate Annual Reports, Inc., in Walter Herdeg, *Graphis/Diagrams* (Zurich, 1976), p. 23.

Puzzles and Hierarchy in Graphics

The complexity of multifunctioning elements can sometimes turn data graphics into visual puzzles, crypto-graphical mysteries for the viewer to decode. A sure sign of a puzzle is that the graphic must be interpreted through a verbal rather than a visual process.

For example, despite its clever and multifunctioning data measure, formed by crossing two four-color grids, this is a puzzle graphic. Deployed here, in a feat of technological virtuosity, are 16 shades of color spread on 3,056 counties, a monument to a sophisticated computer graphics system.[4] But it is surely a graphic experienced verbally, not visually. Over and over, the viewers must run little phrases through their minds, trying to maintain the right pattern of words to make sense out of the visual montage: "Now let's see, purple represents counties where there are both high levels of male cardiovascular disease mortality *and* 11.6 to 56.0 percent of the households have more than 1.01 persons per room. . . . What does that mean anyway? . . . And the yellow-green counties. . . ." By contrast, in a non-puzzle graphic, the translation of visual to verbal is quickly learned, automatic, and implicit —so that the visual image flows right *through* the verbal decoder initially necessary to understand the graphic. As Paul Valéry wrote, "Seeing is forgetting the name of the thing one sees."

[4] The technique is described in Vincent P. Barabba and Alva L. Finkner, "The Utilization of Primary Printing Colors in Displaying More than One Variable," in Bureau of the Census, Technical Paper No. 43, *Graphical Presentation of Statistical Information* (Washington, D.C., 1978), 14–21. The maps are assessed in Howard Wainer and C. M. Francolini, "An Empirical Inquiry Concerning Human Understanding of Two-Variable Color Maps," *American Statistician*, 34 (1980), 81–93.

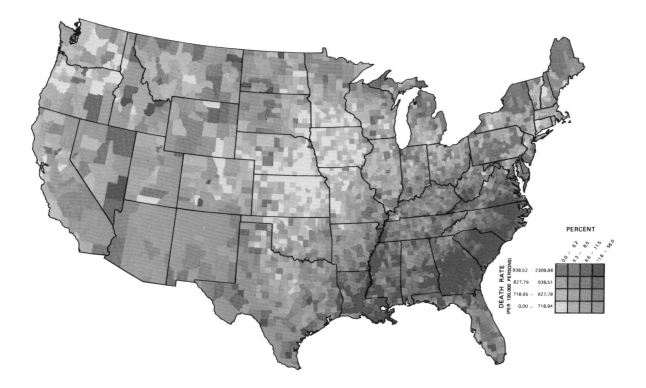

Color often generates graphical puzzles. Despite our experiences with the spectrum in science textbooks and rainbows, the mind's eye does not readily give a visual ordering to colors, except possibly for red to reflect higher levels than other colors, as in the hot spots of the cancer map. Attempts to give colors an order result in those verbal decoders and the mumbling of little mental phrases again—indeed, even mnemonic phrases *about* the phrases required for graphical decoding:

> A method of coloring ingenious in idea but not very satisfactory in practice was used by L. L. Vauthier. It was called the mountain-to-the-sea method. White was used for the representation of the greatest intensity of the fact because it indicated the summit of a mountain with its eternal snow, next came green representing the forests farther down the slopes, then yellow for the grain of the plains, and finally for the minimum the blue of the waters at sea level.[5]

Because they do have a natural visual hierarchy, varying shades of gray show varying quantities better than color. Ten gray shades worked effectively in the galaxies map:

[5] H. Gray Funkhouser, "Historical Development of the Graphical Representation of Statistical Data," *Osiris*, 3 (1937), 326, who cites É. Cheysson, "Les méthodes de statistique graphique à l'Exposition universelle de 1878," *Journal de la Société de Statistique de Paris*, 19 (1878), 331.

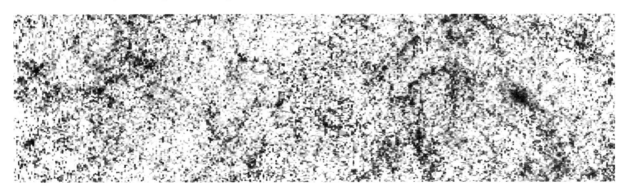

The success of gray compared to the visually more spectacular color gives us a lead on how multifunctioning graphical elements can communicate complex information without turning into puzzles. The shades of gray provide an easily comprehended order to the data measures. This is the key. Central to maintaining clarity in the face of the complex are graphical methods that *organize and order the flow of graphical information* presented to the eye.

How can graphical architecture promote the ordered, sequenced, hierarchical flow of information from the graphic to the mind's eye? How can the data-information be arranged so that the viewer is able to peel back layer after layer of data from a graphic?

Multiple layers of information are created by *multiple viewing depths* and *multiple viewing angles*.

Graphics can be designed to have at least three viewing depths: (1) what is seen from a distance, an overall structure usually aggregated from an underlying microstructure; (2) what is seen up close and in detail, the fine structure of the data; and (3) what is seen implicitly, underlying the graphic—that which is behind the graphic. Look at all the different levels of detail created by this population density map of the United States, a glory of modern cartography prepared by the Bureau of the Census. Each dot, except in urban centers, represents 500 people. Note the corridors connecting the major urban complexes; the effects of landforms on the population distribution (the central valley of California, the valleys and ridges of Appalachia, and the clusters along rivers); and the small towns along the highways, linked like a string of pearls. The map arrays, in effect, some 400,000 points on its implicit grid.

Different visual angles for different aspects of the data also organize graphical information. Each separate line of sight should remain unchanging (preferably horizontal or vertical) as the eye watches for data variation off the flat of the line of sight. For multivariate work, several clear lines can be created. Recall Ayres' display of American divisions in France. Even with its complex, interwoven data, the graphic is not a puzzle. Three separate visual angles make the flow of information coherent: the profile of the horizon for the upward-moving time-series, the vertical for the composition of the bar, and the horizontal for each division's stay. Thus while every drop of ink serves three different data display functions, each of the three comes to the eye with its own independence and integrity.

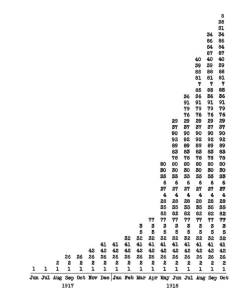

URBAN POPULATION

URBANIZED AREAS

Extent of areas

PLACES OUTSIDE
URBANIZED AREAS

25,000 - 50,000
10,000 - 25,000
2,500 - 10,000

RURAL POPULATION

Places of 1,000—
2,500
Each dot represents
500 of remaining
population

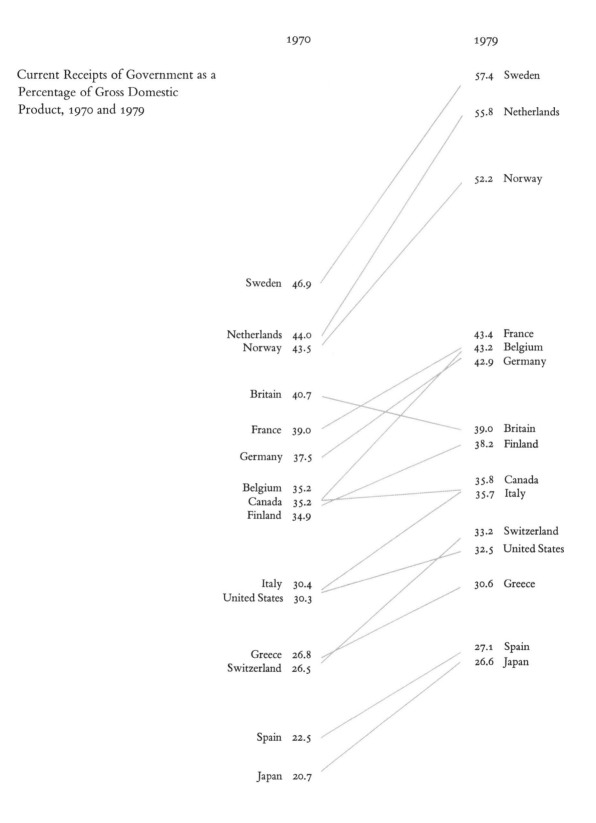

Current Receipts of Government as a
Percentage of Gross Domestic
Product, 1970 and 1979

Similarly, this table-graphic organizes data for viewing in several directions. The chart, when read vertically, ranks 15 countries by government tax collections in 1970 and again in 1979, with the names spaced in proportion to the percentages. Across the columns, the paired comparisons show how the numbers changed over the years. The slopes are also compared by reading down the collection of lines, and lines of unusual slope stand out from the overall upward pattern. The information shown is both integrated and separated: integrated through its connected content, separated in that the eye follows several different and uncluttered paths in looking over the data:

Such an analysis of the *viewing architecture* of a graphic will help in creating and evaluating designs that organize complex information hierarchically.

I want to reach that state of condensation of sensations
which constitutes a picture.

Henri Matisse

Our eyes can make a remarkable number of distinctions within a small area. With the use of very light grid lines, it is easy to locate 625 points in one square inch or, equivalently, 100 points in one square centimeter.

Or consider how an 80 by 80 grid over a square inch—about 30 by 30 over a square centimeter—divides the space:[1]

With the help of considerable redundancy and context, our eyes make fine distinctions of this sort all the time. Measurement instruments used in engineering, architectural, and machine work are engraved with scales of 20 increments to the centimeter and 50 to the inch. Or consider the reading of fine print. The type in the U.S. *Statistical Abstract* is set at 12 lines per vertical inch, with each line running at about 23 characters per inch for a maximum density of 276 characters per square inch. The actual density, given the white space, is in this case 185 characters per square inch or 28 per square centimeter.

25,281 distinctions

[1] A square grid formed on each side by n parallel black and n−1 parallel white lines contains n^2 intersections of two black lines (corners of squares), $(n-1)^2$ intersections of two white lines (white squares), and $2n(n-1)$ intersections of a black and white line (sides of squares), for a total of $(2n-1)^2$ line intersections or distinct locations.

NO. 1450. STEEL PRODUCTS—NET SHIPMENTS, BY MARKET CLASSES: 1960 TO 1978

[In thousands of short tons. Comprises carbon, alloy, and stainless steel. "N.e.c." means not elsewhere classified]

MARKET CLASS	1960	1965	1970	1973	1974	1975	1976	1977	1978
Total [1]	71,149	92,666	90,798	111,430	109,472	79,957	89,447	91,147	97,935
Steel for converting and processing_	2,928	3,932	3,443	4,714	4,486	3,255	4,036	3,679	4,612
Independent forgers, n.e.c._____	841	1,250	1,048	1,213	1,339	1,098	952	998	1,192
Industrial fasteners [2]_____	1,071	1,234	1,005	1,278	1,331	675	912	848	870
Steel service centers, distributors__	11,125	14,813	16,025	20,383	20,400	12,700	14,615	15,346	17,333
Construction, incl. maintenance___	9,664	11,836	8,913	10,731	11,360	8,119	7,508	7,553	9,612
Contractors' products_____	3,602	5,018	4,440	6,459	6,249	3,927	4,502	4,500	3,480
Automotive_____	14,610	20,123	14,475	23,217	18,928	15,214	21,351	21,490	21,253
Rail transportation_____	2,525	3,805	3,098	3,228	3,417	3,152	3,056	3,238	3,549
Freight cars, passenger cars, locomotives_____	1,763	2,875	2,005	1,997	2,097	1,794	1,428	1,709	2,188
Rails and all other [3]_____	762	930	1,093	1,231	1,320	1,358	1,628	1,529	1,361
Shipbuilding and marine equip____	622	1,051	859	1,019	1,339	1,413	969	869	845
Aircraft and aerospace_____	78	94	56	69	79	69	59	63	60
Oil and gas industries_____	1,759	1,936	3,550	3,405	4,210	4,171	2,653	3,650	4,140
Mining, quarrying, and lumbering_	288	392	497	534	644	596	536	486	508
Agricultural, incl. machinery_____	1,003	1,483	1,126	1,772	1,859	1,429	1,784	1,743	1,805
Machinery, industrial equip., tools	3,958	5,873	5,169	6,351	6,440	5,173	5,180	5,566	5,992
Electrical equipment_____	2,078	2,985	2,694	3,348	3,242	2,173	2,671	2,639	2,811
Appliances, utensils, and cutlery___	1,760	2,179	2,160	2,747	2,412	1,653	1,950	2,129	2,094
Other domestic commercial equip_	1,959	2,179	1,778	1,990	1,941	1,390	1,813	1,846	1,889
Containers, packaging, shipping___	6,429	7,331	7,775	7,811	8,218	6,053	6,914	6,714	6,595
Cans and closures_____	4,976	5,867	6,239	6,070	6,349	4,859	5,290	5,173	4,950
Ordnance and other military_____	165	289	1,222	918	654	405	219	193	207
Exports (reporting companies only)	2,563	2,078	5,985	3,138	3,961	1,755	1,839	1,076	1,224

[1] Total includes nonclassified shipments, and, beginning 1970, data include estimates for a relatively small number of companies which report raw steel production but not shipments.   [2] Bolts, nuts, rivets, and screws.   [3] Includes railways, rapid transit systems, railroad rails, trackwork, and equipment.

U.S. Bureau of the Census, *Statistical Abstract of the United States: 1979* (Washington, D.C., 1979), p. 822.

Maps routinely present even finer detail. A cartographer writes that "the resolving power of the eye enables it to differentiate to 0.1 mm where provoked to do so. Clearly, therefore, conciseness is of the essence and high resolution graphics are a common denominator of cartography."[2] Distinctions at 0.1 mm mean 254 per inch.

[2] D. P. Bickmore, "The Relevance of Cartography," in John C. Davis and Michael J. McCullagh, eds., *Display and Analysis of Spatial Data* (London, 1975), p. 331.

How many statistical graphics take advantage of the ability of the eye to detect large amounts of information in small spaces? And how much information should graphics show? Let us begin by considering an empirical measure of graphical performance, the data density.

Data Density in Graphical Practice

The numbers that go into a graphic can be organized into a data matrix of observations by variables. Taking into account the size of the graphic in relation to the amount of data displayed yields the *data density*:

$$\text{data density of a graphic} = \frac{\text{number of entries in data matrix}}{\text{area of data graphic}}$$

Data matrices and data densities vary enormously in practice. At one extreme, this overwrought display (originally printed in five colors) presents a data matrix of four entries, the names and the numbers for the two bars on the right. The left bar is merely the total of the other two. The graph covers 26.5 square inches (171 square centimeters), resulting in a data density of .15 numbers per square inch (.02 numbers per square centimeter), which is thin indeed.

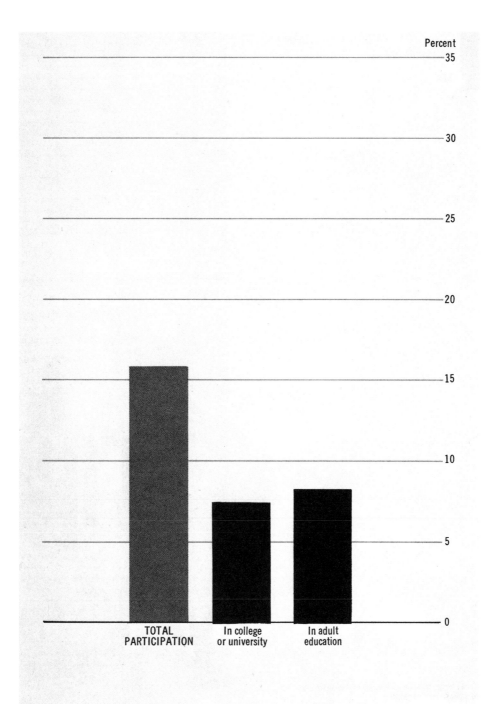

Executive Office of the President, Office
of Management and Budget, *Social
Indicators, 1973* (Washington, D.C.,
1973), p. 86.

The exemplar from the JASA style sheet comes in at a light-weight 3.8 numbers per square inch (0.6 numbers per square centimeter) and a small data matrix of 32 entries:

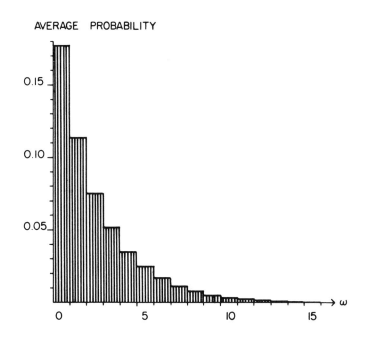

In contrast, the New York weather history, in this reduced version, does very well at 181 numbers per square inch (28 per square centimeter):

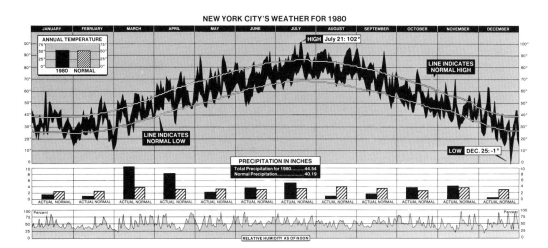

An annual sunshine record reports about 1,000 numbers per square inch (160 per square centimeter):

F. J. Monkhouse and H. R. Wilkinson, *Maps and Diagrams* (London, third edition, 1971), pp. 242–243.

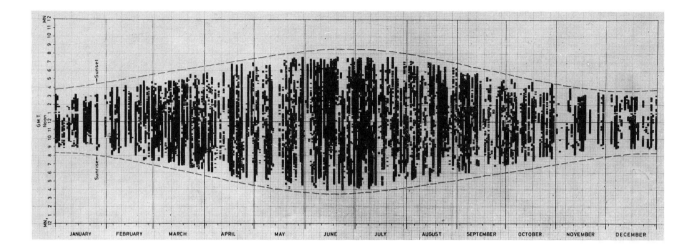

The visual metaphor corresponds appropriately to the data if the image is reversed, so that the light areas are the times when the sun shines:

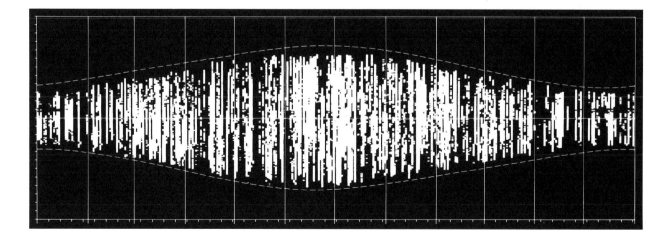

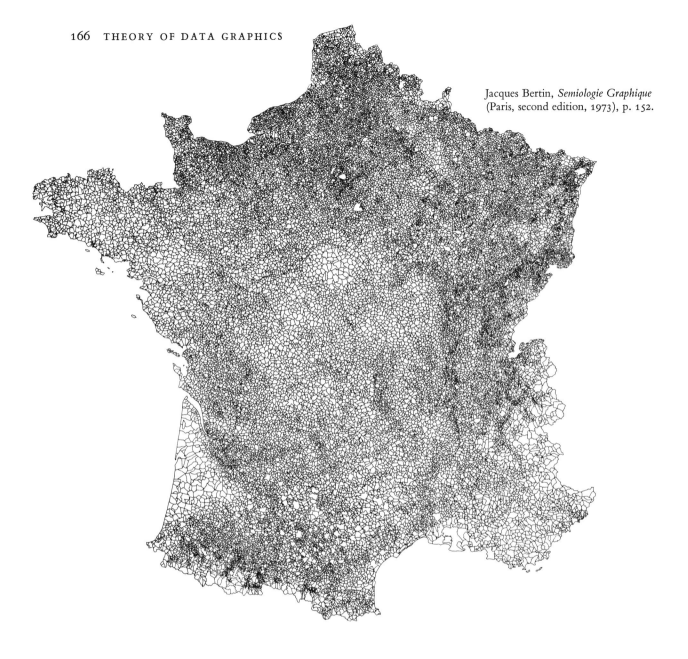

Jacques Bertin, *Semiologie Graphique* (Paris, second edition, 1973), p. 152.

This map (27 square inches, 175 square centimeters) shows the location and boundaries of 30,000 communes of France. It would require at least 240,000 numbers to recreate the data of the map (30,000 latitudes, 30,000 longitudes, and perhaps six numbers describing the shape of each commune). Thus that data density is nearly 9,000 numbers per square inch, or 1,400 numbers per square centimeter.

The new map of the galaxies locates 2,275,328 encoded rectangles on a two-dimensional surface of 61 square inches (390 square centimeters). Each rectangle represents three numbers (two by its location, one by its shading), yielding a data density of 110,000 numbers per square inch or 17,000 numbers per square centimeter. That is the current record.

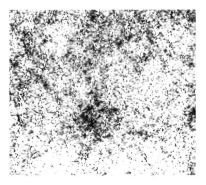

Data Density and the Size of the Data Matrix: Publication Practices

The table shows the data density and the size of the data matrix for graphics sampled from scientific and news publications. At least 20 graphics from each publication were examined.

The table records an enormous diversity of graphical performances both within and between publications. A few data-rich designs appear in nearly every publication. The opportunity is there but it is rarely exploited: the average published graphic is rather thin,

Data Density and Size of Data Matrix,
Statistical Graphics in Selected Publications, Circa 1979–1980

	Data Density (Numbers per square inch) median minimum maximum			Size of Data Matrix median minimum maximum		
Nature	48	3	362	177	15	3780
Journal of the Royal Statistical Society, B	27	4	115	200	10	1460
Science	21	5	44	109	26	316
Wall Street Journal	19	3	154	135	28	788
Fortune	18	5	31	96	42	156
The Times (London)	18	2	122	50	14	440
Journal of the American Statistical Association	17	4	167	150	46	1600
Asahi	13	2	113	29	15	472
New England Journal of Medicine	12	3	923	84	8	3600
The Economist	9	1	51	36	3	192
Le Monde	8	1	17	66	11	312
Psychological Bulletin	8	1	74	46	8	420
Journal of the American Medical Association	7	1	39	53	14	735
New York Times	7	1	13	35	6	580
Business Week	6	2	12	32	14	96
Newsweek	6	1	13	23	2	96
Annuaire Statistique de la France	6	1	25	96	12	540
Scientific American	5	1	69	46	14	652
Statistical Abstract of the United States	5	2	23	38	8	164
American Political Science Review	2	1	10	16	9	40
Pravda	0.2	0.1	1	5	4	20

based on about 50 numbers shown at the rate of 10 per square inch. Among the world's newspapers, the *Wall Street Journal*, *The Times* (London), and *Asahi* publish data-rich graphics, with data densities equal to those of the *Journal of the American Statistical Association*. Most of the American papers and magazines, along with *Pravda*, publish less data per graphic than the major papers of other industrialized countries.

Very few statistical graphics achieve the information display rates found in maps. Highly detailed maps portray 100,000 to 150,000 bits per square inch. For example, the average U.S. Geological Survey topographic quadrangle (measuring 17 by 23 inches) is estimated to contain over 100 million bits of information, or about 250,000 per square inch (40,000 per square centimeter).[3] Perhaps some day statistical graphics will perform as successfully as maps in carrying information.

[3] Morris M. Thompson, *Maps for America* (Washington, D.C., 1979), p. 187.

High-Information Graphics

Data graphics should often be based on large rather than small data matrices and have a high rather than low data density. More information is better than less information, especially when the marginal costs of handling and interpreting additional information are low, as they are for most graphics. The simple things belong in tables or in the text; graphics can give a sense of large and complex data sets that cannot be managed in any other way. If the graphic becomes overcrowded (although several thousand numbers represented may be just fine), a variety of data-reduction techniques—averaging, clustering, smoothing—can thin the numbers out before plotting.[4] Summary graphics can emerge from high-information displays, but there is nowhere to go if we begin with a low-information design.

Data-rich designs give a context and credibility to statistical evidence. Low-information designs are suspect: what is left out, what is hidden, why are we shown so little? High-density graphics help us to compare parts of the data by displaying much information within the view of the eye: we look at one page at a time and the more on the page, the more effective and comparative our eye can be.[5] The principle, then, is:

> Maximize data density and the size of the data
> matrix, within reason.

High-information graphics must be designed with special care. As the volume of data increases, data measures must shrink (smaller dots for scatters, thinner lines for busy time-series). The clutter of

[4] Paul A. Tukey and John W. Tukey, "Summarization; Smoothing; Supplemented Views," in Vic Barnett, ed., *Interpreting Multivariate Data* (Chichester, England, 1982), ch. 12; and William S. Cleveland, "Robust Locally Weighted Regression and Smoothing Scatterplots," *Journal of the American Statistical Association*, 74 (1979), 829–836, are recent papers in the large literature.

[5] It is suggested in the analysis of x-ray films to "search a reduced image so that the whole display can be perceived on at least one occasion without large eye movement." Edward Llewellyn Thomas, "Advice to the Searcher or What Do We Tell Them?" in Richard A. Monty and John W. Senders, eds., *Eye Movements and Psychological Processes* (Hillsdale, N.J., 1976), p. 349.

chartjunk, non-data-ink, and redundant data-ink is even more costly than usual in data-rich designs.

The way to increase data density other than by enlarging the data matrix is to reduce the area of a graphic. The Shrink Principle has wide application:

> Graphics can be shrunk way down.

Many data graphics can be reduced in area to half their currently published size with virtually no loss in legibility and information. For example, Bertin's crisp and elegant line allows the display of 17 small-scale graphics on a single page along with extensive text. Repeated application of the Shrink Principle leads to a powerful and effective graphical design, the small multiple.

Jacques Bertin, *Semiologie Graphique* (Paris, second edition, 1973), p. 214.

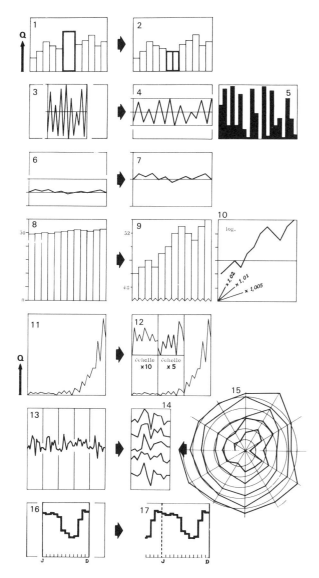

PROBLEMES GRAPHIQUES POSES PAR LES CHRONIQUES

Un total sur deux cases (sur deux ans) doit être divisé par deux (1).
Un total pour six mois sera multiplié par deux dans des cases annuelles.

Courbes trop pointues, réduire l'échelle des Q; la sensibilité angulaire s'inscrit dans une zone moyenne autour de 70°.
Si la courbe n'est pas réductible (grandes et petites variations) employer les colonnes remplies (5).
Courbes trop plates : augmenter l'échelle des Q.

Variations très faibles par rapport au total. Celui-ci perd de l'importance et le zéro peut être supprimé, à condition que le lecteur voit sa suppression (9). Le graphique peut être interprété comme une accélération si l'étude fine des variations est nécessaire (échelle logarithmique (10) (v. p. 240).

Très grande amplitude entre les valeurs extrêmes. Il faut admettre :
1°) Soit de ne pas percevoir les plus petites variations.
2°) Soit de ne s'intéresser qu'aux différences relatives (échelle logarithmique) sans connaître la quantité absolue.
3°) Soit admettre des périodes différentes dans la composante ordonnée et les traiter à des échelles différentes au-dessus de l'échelle commune (12).

Cycles très marqués.
Si l'étude porte sur la comparaison des phases de chaque cycle, il est préférable de décomposer (13) de manière à superposer les cycles (14). La construction polaire peut être employée, de préférence dans une forme spirale (15) (ne pas commencer par un trop petit cercle); pour spectaculaire qu'elle soit, elle est moins efficace que la construction orthogonale.

Courbes annuelles de pluie ou de température. Un cycle possède deux phases (17), pourquoi n'en offrir qu'une à la perception du spectateur ? (16).

Small Multiples

Small multiples resemble the frames of a movie: a series of graphics, showing the same combination of variables, indexed by changes in another variable. Twenty-three hours of Los Angeles air pollution are organized into this display, based on a computer generated video tape. Shown is the hourly average distribution of reactive hydrocarbon emissions. The design remains constant through all the frames, so that attention is devoted entirely to shifts in the data:

From video tape by Gregory J. McRae, California Institute of Technology. The model is described in G. J. McRae, W. R. Goodin, and J. H. Seinfeld, "Development of a Second-Generation Mathematical Model for Urban Air Pollution. I. Model Formulation," *Atmospheric Environment*, 16 (1982), 679–696.

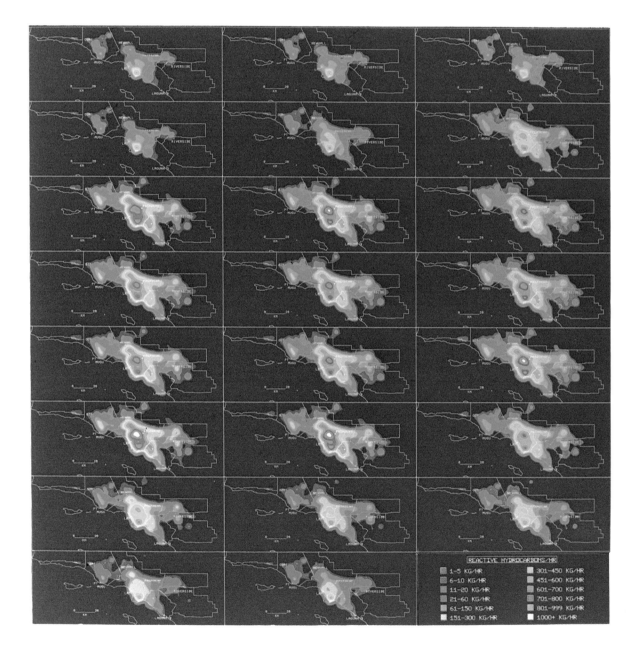

These grim small multiples show the distribution of occurrence of the cancer melanoma. The sites of 269 primary melanomas are recorded, along with the distribution between men and women. Note the data graphical arithmetic, similar to that of the multiwindow plot.

Arthur Wiskemann, "Zur Melanomentstehung durch chronische Lichteinwirkung," *Der Hautarzt*, 25 (1974), 21.

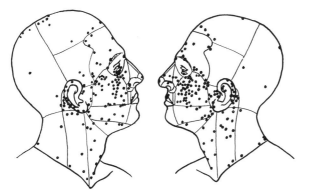

Abb. 1. Verteilung von 269 primären Melanomen auf Kopf und Hals

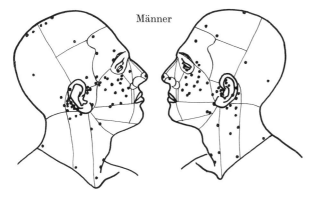

Männer

Abb. 2

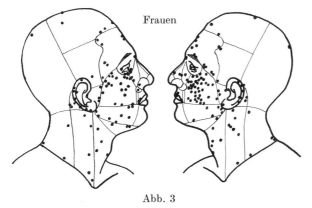

Frauen

Abb. 3

Abb. 2 u. 3. Differenzierung der Melanomverteilung nach Geschlechtern

The effects of sampling errors are shown in these 12 distributions, each based on a sample of 50 random normal deviates:

Edmond A. Murphy, "One Cause? Many Causes? The Argument from the Bimodal Distribution," *Journal of Chronic Diseases*, 17 (1964), 309.

These six distributions show the age composition of herring catches each year from 1908 to 1913. A tremendous number of herring were spawned in 1904, and that class began to dominate the 1908 catch as four-year-olds, then the 1909 catch as five-year-olds, and so on:

Johan Hjort, "Fluctuations in the Great Fisheries of Northern Europe," *Rapports et Proces-Verbaux*, 20 (1914), in Susan Schlee, *The Edge of an Unfamiliar World* (New York, 1973), p. 226.

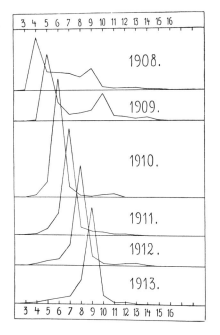

This next design compares a complex set of data: shown are the chromosomes of (from left to right) man, chimpanzee, gorilla, and orangutan. The similarities between humans and the great apes are to be noted.

Jorge J. Yunis and Om Prakash, "The Origin of Man: A Chromosomal Pictorial Legacy," *Science*, 215 (March 19, 1982), 1527.

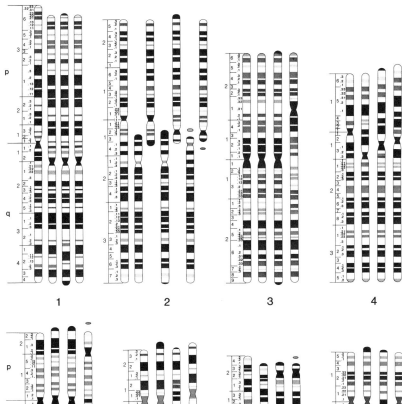

1

2

3

4

5

6

7

8

9

10

11

12

13

14

15

16

17

X

18

19

20

21

22

Y

And, finally, a visually similar small multiple, the *Consumer Reports* frequency-of-repair records for automobiles built from 1976 to 1981. This is a particularly ingenious mix of table and graphic, portraying a complex set of comparisons between manufacturers, types of cars, year, and trouble spots.

Consumer Reports, 47 (April 1982), 199–207. Redrawn.

○ = Much better than average ◦ = Better than average ◯ = Average ◉ = Worse than average ● = Much worse than average

Trouble Spots

Air-conditioning
Body exterior (paint)
Body exterior (rust)
Body hardware
Body integrity
Brakes
Clutch
Driveline
Electrical system (chassis)
Engine cooling
Engine mechanical
Exhaust system
Fuel system
Ignition system
Suspension
Transmission (manual)
Transmission (automatic)
Trouble Index
Cost Index

First row of panels (years 76 77 78 79 80 81):
- Chevrolet Malibu, Chevelle 6, V6
- Chevrolet Monza 4
- Datsun 210, B210
- Ford Granada 6
- Ford pickup truck 6 (2WD)
- Honda Accord

Second row of panels (years 76 77 78 79 80 81):
- Mercedes-Benz 300D 5 (diesel)
- Plymouth Volare 6
- Subaru (except 4WD)
- Toyota Corolla (except Tercel)
- Volkswagen Rabbit (diesel)
- Volvo 240 series

Conclusion

Well-designed small multiples are

- inevitably comparative

- deftly multivariate

- shrunken, high-density graphics

- usually based on a large data matrix

- drawn almost entirely with data-ink

- efficient in interpretation

- often narrative in content, showing shifts in the relationship between variables as the index variable changes (thereby revealing interaction or multiplicative effects).

Small multiples reflect much of the theory of data graphics:

For non-data-ink, less is more.

For data-ink, less is a bore.[6]

[6] The two aphorisms on the meaning of "less" are, respectively, credited to Ludwig Mies van der Rohe and to Robert Venturi, *Complexity and Contradiction in Architecture* (New York, second edition, 1977), p. 17.

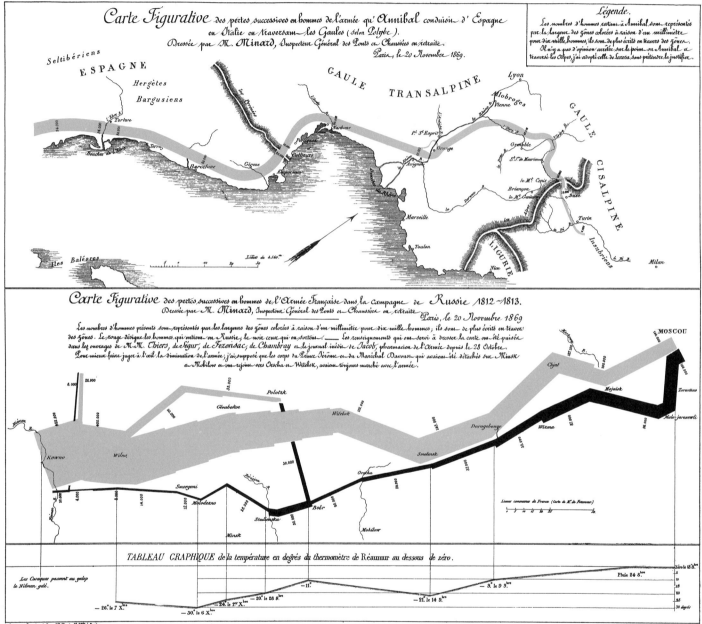

Minard drew at least two versions of Napoleon's march to Moscow, the second in color with additional text describing the data
sources. Another "Carte Figurative" was added in the 1869 plate,
this one portraying Hannibal's campaign in Spain, Gaul, and
Northern Italy. Minard's refined use of color contrasts with the
brutal tones often seen in current-day graphics.

Charles Joseph Minard, *Tableaux Graphiques et Cartes Figuratives de M. Minard,
1845–1869*, a portfolio of his work held
by the Bibliothèque de l'École Nationale
des Ponts et Chaussées, Paris.

What makes for such graphical elegance? What accounts for
the quality of Minard's graphics, of those of Playfair and Marey,
and of some recent work, such as the new view of the galaxies?
Good design has two key elements:

> Graphical elegance is often found in
> simplicity of design and complexity of data.

Visually attractive graphics also gather their power from content and interpretations beyond the immediate display of some
numbers. The best graphics are about the useful and important,
about life and death, about the universe. Beautiful graphics do
not traffic with the trivial.

On rare occasions graphical architecture combines with the data
content to yield a uniquely spectacular graphic. Such performances
can be described and admired but there are no compositional
principles on how to create that one wonderful graphic in a million. As Ben Shahn once said, "Aesthetics is for the artist like
ornithology is for the birds."

What can be suggested, though, are some guides for enhancing
the visual quality of the more routine, workaday designs. Attractive displays of statistical information

- have a properly chosen format and design

- use words, numbers, and drawing together

- reflect a balance, a proportion, a sense of relevant scale

- display an accessible complexity of detail

- often have a narrative quality, a story to tell about the data

- are drawn in a professional manner, with the technical details
  of production done with care

- avoid content-free decoration, including chartjunk.

The Choice of Design: Sentences, Text-Tables, Tables, Semi-Graphics, and Graphics

The substantive content, extensiveness of labels, and volume and ordering of data all help determine the choice of method for the display of quantitative materials. The basic structures for showing data are the sentence, the table, and the graphic. Often two or three of these devices should be combined.

The conventional sentence is a poor way to show more than two numbers because it prevents comparisons within the data. The linearly organized flow of words, folded over at arbitrary points (decided not by content but by the happenstance of column width), offers less than one effective dimension for organizing the data. Instead of:

> Nearly 53 percent of the type A group did something or other compared to 46 percent of B and slightly more than 57 percent of C.

Arrange the type to facilitate comparisons, as in this *text-table*:

> The three groups differed in how they did something or other:
>
> | | |
> |---|---|
> | Group A | 53% |
> | Group B | 46% |
> | Group C | 57% |

There are nearly always better sequences than alphabetical—for example, ordering by content or by data values:

> | | |
> |---|---|
> | Group B | 46% |
> | Group A | 53% |
> | Group C | 57% |

Tables are clearly the best way to show exact numerical values, although the entries can also be arranged in semi-graphical form. Tables are preferable to graphics for many small data sets.[1] A table is nearly always better than a dumb pie chart; the only worse design than a pie chart is several of them, for then the viewer is asked to compare quantities located in spatial disarray both within and between pies, as in this heavily encoded example from an atlas. Given their low data-density and failure to order numbers along a visual dimension, pie charts should never be used.[2]

Department of Surveys, Ministry of Labour, *Atlas of Israel* (Jerusalem, 1956–), vol. 8, p. 8.

[1] On the design of tables, see A.S.C. Ehrenberg, "Rudiments of Numeracy," *Journal of the Royal Statistical Society*, A, 140 (1977), 277–297.

[2] This point is made decisively in Jacques Bertin, *Graphics and Graphic Information Processing* (Berlin, 1981). Bertin describes multiple pie charts as "completely useless" (p. 111).

Tables also work well when the data presentation requires many localized comparisons. In this 410-number table that I designed for the *New York Times* to show how different people voted in presidential elections in the United States, comparisons between the elections of 1980 and 1976 are read across each line; within-election analysis is conducted by reading downward in the clusters of three to seven lines. The horizontal rules divide the data into topical paragraphs; the rows are ordered so as to tell an ordered story about the elections. This type of elaborate table, a *supertable*, is likely to attract and intrigue readers through its organized, sequential detail and reference-like quality. One supertable is far better than a hundred little bar charts.

New York Times, November 9, 1980,
p. A–28.

How Different Groups Voted for President

Based on 12,782 interviews with voters at their polling places. Shown is how each group divided its vote for President and, in parentheses, the percentage of the electorate belonging to each group.

	CARTER	REAGAN	ANDERSON	CARTER-FORD in 1976
Democrats (43%)	66	26	6	77 - 22
Independents (23%)	30	54	12	43 - 54
Republicans (28%)	11	84	4	9 - 90
Liberals (17%)	57	27	11	70 - 26
Moderates (46%)	42	48	8	51 - 48
Conservatives (28%)	23	71	4	29 - 70
Liberal Democrats (9%)	70	14	13	86 - 12
Moderate Democrats (22%)	66	28	6	77 - 22
Conservative Democrats (8%)	53	41	4	64 - 35
Politically active Democrats (3%)	72	19	8	—
Democrats favoring Kennedy in primaries (13%)	66	24	8	—
Liberal Independents (4%)	50	29	15	64 - 29
Moderate Independents (12%)	31	53	13	45 - 53
Conservative Independents (7%)	22	69	6	26 - 72
Liberal Republicans (2%)	25	66	9	17 - 82
Moderate Republicans (11%)	13	81	5	11 - 88
Conservative Republicans (12%)	6	91	2	6 - 93
Politically active Republicans (2%)	5	89	6	—
East (32%)	43	47	8	51 - 47
South (27%)	44	51	3	54 - 45
Midwest (20%)	41	51	6	48 - 50
West (11%)	35	52	10	46 - 51
Blacks (10%)	82	14	3	82 - 16
Hispanics (2%)	54	36	7	75 - 24
Whites (88%)	36	55	8	47 - 52
Female (49%)	45	46	7	50 - 48
Male (51%)	37	54	7	50 - 48
Female, favors equal rights amendment (22%)	54	32	11	—
Female, opposes equal rights amendment (15%)	29	66	4	—
Catholic (25%)	40	51	7	54 - 44
Jewish (5%)	45	39	14	64 - 34
Protestant (46%)	37	56	6	44 - 55
Born-again white Protestant (17%)	34	61	4	—
18 - 21 years old (6%)	44	43	11	48 - 50
22 - 29 years old (17%)	43	43	11	51 - 46
30 - 44 years old (31%)	37	54	7	49 - 49
45 - 59 years old (23%)	39	55	6	47 - 52
60 years or older (18%)	40	54	4	47 - 52
Family income				
Less than $10,000 (13%)	50	41	6	58 - 40
$10,000 - $14,999 (14%)	47	42	8	55 - 43
$15,000 - $24,999 (30%)	38	53	7	48 - 50
$25,000 - $50,000 (24%)	32	58	8	36 - 62
Over $50,000 (5%)	25	65	8	—
Professional or manager (40%)	33	56	9	41 - 57
Clerical, sales or other white-collar (11%)	42	48	8	46 - 53
Blue-collar worker (17%)	46	47	5	57 - 41
Agriculture (3%)	29	66	3	—
Looking for work (3%)	55	35	7	65 - 34
Education				
High school or less (39%)	46	48	4	57 - 43
Some college (28%)	35	55	8	51 - 49
College graduate (27%)	35	51	11	45 - 55
Labor union household (26%)	47	44	7	59 - 39
No member of household in union (62%)	35	55	8	43 - 55
Family finances				
Better off than a year ago (16%)	53	37	8	30 - 70
Same (40%)	46	46	7	51 - 49
Worse off than a year ago (34%)	25	64	8	77 - 23
Family finances and political party				
Democrats, better off than a year ago (7%)	77	16	6	69 - 31
Democrats, worse off than a year ago (13%)	47	39	10	94 - 6
Independents, better off (3%)	45	36	12	—
Independents, worse off (9%)	21	65	11	—
Republicans, better off (4%)	18	77	5	3 - 97
Republicans, worse off (11%)	6	89	4	24 - 76
More important problem				
Unemployment (39%)	51	40	7	75 - 25
Inflation (44%)	30	60	9	35 - 65
Feel that U.S. should be more forceful in dealing with Soviet Union even if it would increase the risk of war (54%)	28	64	6	—
Disagree (31%)	56	32	10	—
Favor equal rights amendment (46%)	49	38	11	—
Oppose equal rights amendment (35%)	26	68	4	—
When decided about choice				
Knew all along (41%)	47	50	2	44 - 55
During the primaries (13%)	30	60	8	57 - 42
During conventions (8%)	36	55	7	51 - 48
Since Labor Day (8%)	30	54	13	49 - 49
In week before election (23%)	38	46	13	49 - 47

Source: 1976 and 1980 election day surveys by The New York Times/CBS News Poll and 1976 election day survey by NBC News.

For sets of highly labeled numbers, a wordy data graphic—
coming close to straight text—works well. This table of numbers
is nicely organized into a graphic:

Some Winners and Losers in the Forecasting Game

About a year ago, eight forecasters were asked for their predictions on some key economic indicators. Here's how the forecasts stack up against the probable 1978 results (shown in the black panel).

Council of Economic Advisers: +4.7%
Data Resources: +4.5%
Nat. Assoc. of Business Economists: +4.5%
Wharton Econometric Forecasting: +4.5%
Congressional Budget Office: +4.4%
Conference Board: +4.2%
I.B.M. Economics Department: +4.1%

Nat. Assoc. of Business Economists: +6.2%
I.B.M. Economics Department: +5.9%

Wharton Econometric Forecasting: +21%

Chase Econometrics: 7.4%
Wharton Econometric Forecasting: 6.8%
Conference Board: 6.7%
Nat. Assoc. of Business Economists: 6.7%
I.B.M. Economics Department: 6.6%
Data Resources: 6.5%
Congressional Budget Office: 6.3%
Council of Economic Advisers: 6.3%

| Real G.N.P. Growth: +3.8% | Industrial Production Growth: +5.8% | Change in Consumer Prices: +7.7% | Corporate Profits Growth: +13.3% | Unemployment Rate: 6% |

Chase Econometrics: +2.8%

Conference Board: +5.5%
Data Resources: +5.2%
Wharton Econometric Forecasting: +4.8%
Chase Econometrics: +1.9%

I.B.M. Economics Department: +6.6%
Nat. Assoc. of Business Economists: +6.5%
Conference Board: +6.2%
Data Resources: +6.2%
Chase Econometrics: +5.9%
Council of Economic Advisers: +5.9%
Wharton Econometric Forecasting: +5.4%

Data Resources: +10.5%
I.B.M. Economics Department: +10.4%
Chase Econometrics: +6.5%

Forecasters are not listed in categories for which they did not make a prediction.
*After taxes

New York Times, January 2, 1979, p. D-3.

Making Complexity Accessible: Combining Words, Numbers, and Pictures

Explanations that give access to the richness of the data make
graphics more attractive to the viewer. Words and pictures are
sometimes jurisdictional enemies, as artists feud with writers for
scarce space. An unfortunate legacy of these craft-union differences
is the artificial separation of words and pictures; a few style sheets
even forbid printing on graphics. What has gone wrong is that the
techniques of production instead of the information conveyed
have been given precedence.

Words and pictures belong together. Viewers need the help that
words can provide. Words on graphics are data-ink, making
effective use of the space freed up by erasing redundant and non-
data-ink. It is nearly always helpful to write little messages on the
plotting field to explain the data, to label outliers and interesting
data points, to write equations and sometimes tables on the graphic
itself, and to integrate the caption and legend into the design so
that the eye is not required to dart back and forth between textual
material and the graphic. (The size of type on and around graphics

can be quite small, since the phrases and sentences are usually not too long—and therefore the small type will not fatigue viewers the way it does in lengthy texts.)

The principle of *data/text integration* is

> Data graphics are paragraphs about data and
> should be treated as such.

Words, graphics, and tables are different mechanisms with but a single purpose—the presentation of information. Why should the flow of information be broken up into different places on the page because the information is packaged one way or another? Sometimes it may be useful to have multiple story-lines or multiple levels of presentation, but that should be a deliberate design judgment, not something decided by conventional production requirements. Imagine if graphics were replaced by paragraphs of words and those paragraphs scattered over the pages out of sequence with the rest of the text—that is how graphical and tabular information is now treated in the layout of many published pages, particularly in scientific journals and professional books.

Tables and graphics should be run into the text whenever possible, avoiding the clumsy and diverting segregation of "See Fig. 2," (figures all too often located on the back of the adjacent page).[3] If a display is discussed in various parts of the text, it might well be printed afresh near each reference to it, perhaps in reduced size in later showings. The principle of text/graphic/table integration also suggests that the same typeface be used for text and graphic and, further, that ruled lines separating different types of information be avoided. Albert Biderman notes that illustrations were once well-integrated with text in scientific manuscripts, such as those of Newton and Leonardo da Vinci, but that statistical graphics became segregated from text and table as printing technology developed:

> The evolution of graphic methods as an element of the scientific enterprise has been handicapped by their adjunctive, segregated, and marginal position. The exigencies of typography that moved graphics to a segregated position in the printed work have in the past contributed to their intellectual segregation and marginality as well. There was a corresponding organizational segregation, with decisions on graphics often passing out of the hands of the original analyst and communicator into those of graphic specialists—the commercial artists and designers of graphic departments and audio-visual aids shops, for example, whose predilections and skills are usually more those of cosmeticians and merchandisers than of scientific analysts and communicators.[4]

[3] "Fig.," often used to refer to graphics, is an ugly abbreviation and is not worth the two spaces saved.

[4] Albert D. Biderman, "The Graph as a Victim of Adverse Discrimination and Segregation," *Information Design Journal*, 1 (1980), 238.

Page after page of Leonardo's manuscripts have a gentle but thorough integration of text and figure, a quality rarely seen in modern work:

Leonardo da Vinci, *Treatise on Painting* [*Codex Urbinas Latinus 1270*], vol. 2, facsimile (Princeton, 1956), p. 234, paragraph 827.

Finally, a caveat: the use of words and pictures together requires a special sensitivity to the purpose of the design—in particular, whether the graphic is primarily for communication and illustration of a settled finding or, in contrast, for the exploration of a data set. Words on and around graphics are highly effective— sometimes all too effective—in telling viewers how to allocate their attention to the various parts of the data display.[5] Thus, for graphics in exploratory data analysis, words should tell the viewer *how* to read the design (if it is a technically complex arrangement) and not *what* to read in terms of content.

[5] Experiments in visual perception indicate that word instructions substantially determine eye movements in viewing pictures. See John D. Gould, "Looking at Pictures," in Richard A. Monty and John W. Senders, eds., *Eye Movements and Psychological Processes* (Hillsdale, N.J., 1976), 323–343.

Accessible Complexity: The Friendly Data Graphic

An occasional data graphic displays such care in design that it is particularly accessible and open to the eye, as if the designer had the viewer in mind at every turn while constructing the graphic. This is the *friendly data graphic*.

There are many specific differences between friendly and unfriendly graphics:

Friendly	Unfriendly
words are spelled out, mysterious and elaborate encoding avoided	abbreviations abound, requiring the viewer to sort through text to decode abbreviations
words run from left to right, the usual direction for reading occidental languages	words run vertically, particularly along the Y-axis; words run in several different directions
little messages help explain data	graphic is cryptic, requires repeated references to scattered text
elaborately encoded shadings, cross-hatching, and colors are avoided; instead, labels are placed on the graphic itself; no legend is required	obscure codings require going back and forth between legend and graphic
graphic attracts viewer, provokes curiosity	graphic is repellent, filled with chartjunk
colors, if used, are chosen so that the color-deficient and color-blind (5 to 10 percent of viewers) can make sense of the graphic (blue can be distinguished from other colors by most color-deficient people)	design insensitive to color-deficient viewers; red and green used for essential contrasts
type is clear, precise, modest; lettering may be done by hand	type is clotted, overbearing
type is upper-and-lower case, with serifs	type is all capitals, sans serif

With regard to typography, Josef Albers writes:

> The concept that "the simpler the form of a letter the simpler its reading" was an obsession of beginning constructivism. It became something like a dogma, and is still followed by "modernistic" typographers. . . . Ophthalmology has disclosed that the more the letters are differentiated from each other, the easier is the reading. Without going into comparisons and details, it should be realized that words consisting of only capital letters present the most difficult reading—because of their equal height, equal volume, and, with most, their equal width. When comparing serif letters with sans-serif, the latter provide an uneasy reading. The fashionable preference for sans-serif in text shows neither historical nor practical competence.[6]

[6] Josef Albers, *Interaction of Color* (New Haven, 1963, revised edition 1975), p. 4.

Proportion and Scale: Line Weight and Lettering

Graphical elements look better together when their relative proportions are in balance. An integrated quality, an appropriate visual linkage between the various elements, results. This musical score of Karlheinz Stockhausen exhibits such a visual balance:

Karlheinz Stockhausen, *Texte*, vol. 2 (Cologne, 1964), p. 82, from the score of "Zyklus für einen Schlagzeuger."

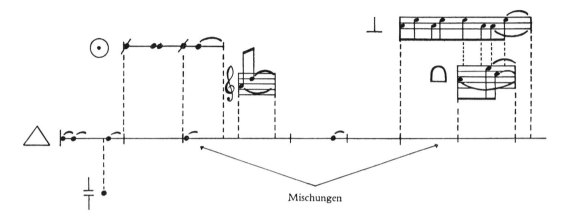

Mischungen

In contrast, this next design is heavy handed, with nearly every element out of balance: the clotted ink, the poor style of lettering, the puffed-up display of a small data set, the coarse texture of the entire graphic, and the mismatch between drawing and surrounding text:

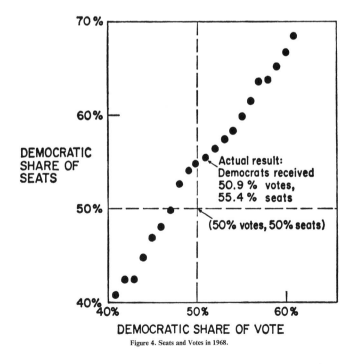

Figure 4. Seats and Votes in 1968.

Edward R. Tufte, "The Relationship Between Seats and Votes in Two-Party Systems," *American Political Science Review*, 67 (June 1973), 551.

Lines in data graphics should be thin. One reason eighteenth-
and nineteenth-century graphics look so good is that they were
engraved on copper plates, with a characteristic hair-thin line.
The drafting pens of twentieth-century mechanical drawing
thickened linework, making it clumsy and unattractive.

An effective aesthetic device is the orthogonal intersection of
lines of different weights:

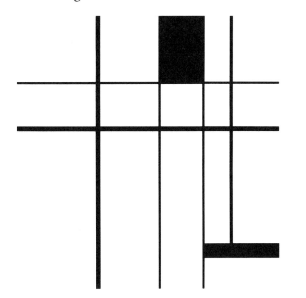

Poster for the exhibition "Mondrian and
Neo-Plasticism in America," Yale Uni-
versity Art Gallery, October 18 to
December 2, 1979. The original painting
was done in 1941 by Diller; see Nancy
J. Troy, *Mondrian and Neo-Plasticism in
America* (New Haven, 1979), p. 28.

Nearly every intersection of the lines in this design (based on a
painting by Burgoyne Diller) involves lines of differing weights,
and it makes a difference, for the painting's character is diluted
with lines of constant width:

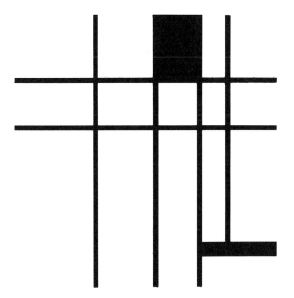

Likewise, data graphics can be enhanced by the perpendicular intersections of lines of differing weights. The heavier line should be a data measure. In a time-series, for example:

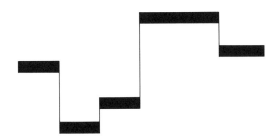

The contrast in line weight represents contrast in meaning. The greater meaning is given to the greater line weight; thus the data line should receive greater weight than the connecting verticals. The logic here is a restatement, in different language, of the principle of data-ink maximization.

Proportion and Scale: The Shape of Graphics

Graphics should tend toward the horizontal, greater in length than height:

Several lines of reasoning favor horizontal over vertical displays.
   First, analogy to the horizon. Our eye is naturally practiced in detecting deviations from the horizon, and graphic design should take advantage of this fact. Horizontally stretched time-series are more accessible to the eye:

The analogy to the horizon also suggests that a shaded, high contrast display might occasionally be better than the floating snake.
The shading should be calm, without moiré effects.

Second, ease of labeling. It is easier to write and to read words
that read from left to right on a horizontally stretched plotting-
field:

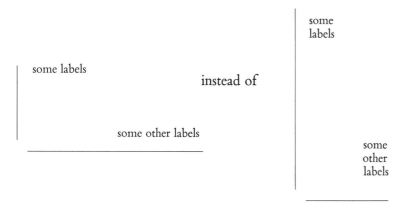

Third, emphasis on causal influence. Many graphics plot, in essence,

and a longer horizontal helps to elaborate the workings of the
causal variable in more detail.

Fourth, Tukey's counsel.

> Most diagnostic plots involve either a more or less definite dependence that bobbles around a lot, or a point spatter. Such plots are rather more often better made *wider* than tall. Wider-than-tall shapes usually make it easier for the eye to follow from left to right.
>
> Perhaps the most general guidance we can offer is that smoothly-changing curves can stand being taller than wide, but a wiggly curve needs to be wider than tall. . . .[7]

[7]John W. Tukey, *Exploratory Data Analysis* (Reading, Mass., 1977), p. 129.

And, finally, Playfair's example. Of the 89 graphics in six different books by William Playfair, most (92 percent) are wider than tall. Several of the exceptions are his skyrocketing government debt displays. This plot shows the dimensions of each of those 89 graphics:

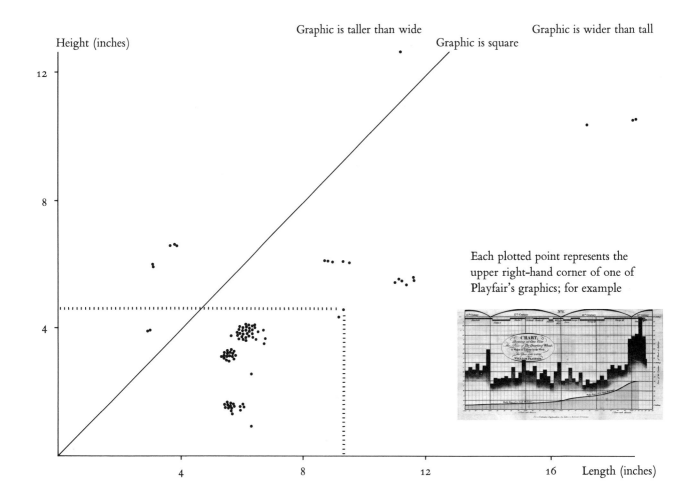

Each plotted point represents the upper right-hand corner of one of Playfair's graphics; for example

If graphics should tend toward the horizontal rather than the vertical, then how much so? A venerable (fifth-century B.C.) but dubious rule of aesthetic proportion is the Golden Section, a "divine division" of a line.[8] A length is divided such that the smaller is to the greater part as the greater is to the whole:

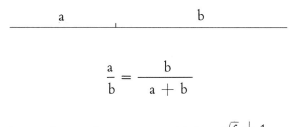

$$\frac{a}{b} = \frac{b}{a+b}$$

Solving the quadratic when a = 1 yields $b = \frac{\sqrt{5}+1}{2} = 1.618\ldots$

In turn the Golden Rectangle is

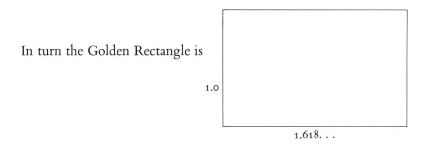

1.0

1.618. . .

The nice geometry of the Golden Rectangle is not unique; Birkhoff points out that at least five other rectangles (including the square) have one simple mathematical property or another for which aesthetic claims might be made:[9]

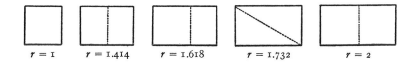

$r = 1$ $r = 1.414$ $r = 1.618$ $r = 1.732$ $r = 2$

Playfair favored proportions between 1.4 and 1.8 in about two-thirds of his published graphics, with most of the exceptions moving more toward the horizontal than the golden prescription:

[8] The combination of geometry and mysticism surrounding the Golden Rectangle can be seen in Miloutine Borissavlièvitch, *The Golden Number and the Scientific Aesthetics of Architecture* (New York, 1958) and Tons Brunés, *The Secrets of Ancient Geometry* (Copenhagen, 1967), vols. 1 and 2.

[9] George D. Birkhoff, *Aesthetic Measure* (Cambridge, 1933), pp. 27–30.

Golden Rectangle

Visual preferences for rectangular proportions have been studied by psychologists since 1860, but, even given the implausible assumption that such studies are relevant to graphic design, the findings are hardly decisive. A mild preference for proportions near to the Golden Rectangle is found among those taking part in the experiments, but the preferred height/length ratios also vary a great deal, ranging between

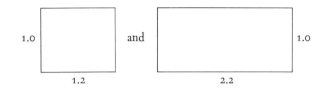

And, as is nearly always the case in experiments in graphical perception, viewer responses were found to be highly context-dependent.[10]

[10] I have relied on Leonard Zusne, *Visual Perception of Form* (New York, 1970), ch. 10, for a summary of the immense literature.

The conclusions:

• If the nature of the data suggests the shape of the graphic, follow that suggestion.

• Otherwise, move toward horizontal graphics about 50 percent wider than tall:

Epilogue: Designs for the Display of Information

Design is choice. The theory of the visual display of quantitative information consists of principles that generate design options and that guide choices among options. The principles should not be applied rigidly or in a peevish spirit; they are not logically or mathematically certain; and it is better to violate any principle than to place graceless or inelegant marks on paper. Most principles of design should be greeted with some skepticism, for word authority can dominate our vision, and we may come to see only through the lenses of word authority rather than with our own eyes.

What is to be sought in designs for the display of information is the clear portrayal of complexity. Not the complication of the simple; rather the task of the designer is to give visual access to the subtle and the difficult—that is,

the revelation of the complex.

Index